Beaches of the Gulf Coast

Harte Research Institute for Gulf of Mexico Studies Series

Sponsored by the Harte Research Institute for Gulf of Mexico Studies, Texas A&M University–Corpus Christi

John W. Tunnell Jr., General Editor

A list of titles in this series appears at the end of the book.

BEACHES OF THE GULF COAST

Richard A. Davis Jr.

TEXAS A&M UNIVERSITY PRESS

College Station

Copyright © 2014 by Richard A. Davis Jr.
All rights reserved
First edition

Manufactured in China by Everbest Printing Co.,
through FCI Print Group
This paper meets the requirements of ANSI/NISO Z39.48–1992
(Permanence of Paper).
Binding materials have been chosen for durability.

LIBRARY OF CONGRESS CATALOGING-IN-PUBLICATION DATA

Davis, Richard A., Jr., 1937–
  Beaches of the Gulf Coast / Richard A. Davis Jr. — 1st ed.
    p. cm. — (Harte Research Institute for Gulf of Mexico Studies series)
  Includes bibliographical references and index.
  "Sponsored by the Harte Research Institute for Gulf of Mexico Studies, Texas A&M University–Corpus Christi."
  ISBN 978-1-62349-038-6 (flex : alk. paper) —
  ISBN 978-1-62349-112-3 (e-book)
  1. Beaches—Gulf Coast (U.S.) 2. Coast changes—Gulf Coast (U.S.) 3. Coastal ecology—Gulf Coast (U.S.) 4. Coastal zone management—Gulf Coast (U.S.) 5. Coasts—Mexico, Gulf of. 6. Gulf Coast (U.S.) I. Title.
  GB459.4.D38 2014
  551.45'7—dc23
  2013022039

Contents

Preface   vii

PART I   **GENERAL CHARACTERISTICS AND DYNAMICS OF BEACHES**

    1. Coastal Processes 3
    2. Beach Geomorphology and Barrier Island Morphodynamics 21
    3. Beach Materials, Structures, and Sources 47
    4. Human Impact on Gulf Beaches 65
    5. Common Animals and Plants of the Gulf Beaches and Surf Zone 91

PART II   **BEACHES ALONG THE GULF OF MEXICO COAST**

    6. Beaches of Florida 105
    7. Beaches of Alabama 143
    8. Beaches of Mississippi 153
    9. Beaches of Louisiana 167
    10. Beaches of Texas 179
    11. Beaches of Mexico and Cuba 209

Glossary   225
Index   229

Preface

MOST of the world's population lives within 50 km of the coast, and the most popular tourist destinations in the world are beaches. The Gulf of Mexico coast is no exception. Several million people live within the coastal zone of this region, and these residents, along with tourists and seasonal visitors, flock to the Gulf Coast to take advantage of the wonderful beaches, adjacent environments, and weather.

This book is designed to provide the interested public with a primer on beaches in general and more specifically on those of the Gulf of Mexico coast. The entire Gulf Coast is covered: from the Florida Keys across the United States and Mexico to Varadero Beach in Cuba. Formation of beaches, beach characteristics, and beach dynamics provide the reader with a good understanding of the nature of this coastal environment—how it behaves and what causes changes to the beach. Given the numerous population centers near beaches and the extensive development associated with the coast, human intervention into the natural beach environment will be discussed and compared with the natural situation. There are various plants and animals that live on and adjacent to beaches, and they are included. Although emphasis is on the beach itself, the adjacent surf zone and dune environments are considered as they relate to the beach.

The geographic coverage of Gulf Coast beaches is similar to a travelogue, following the coast of each state in geographic order and position. Special features are emphasized, and special attention is directed to places where guests may visit. The discussion also includes the reconstruction of eroded beaches by adding sand, called beach nourishment. The numerous photographs are intended to acquaint the reader with the actual nature of the coast at various locations. These photographs show that some places along the Gulf Coast have severe problems, but some beaches are doing very well; some have relied on nourishment for this status, and others have achieved it naturally. The vast majority of the photographs of the US coast, except the sea islands of Mississippi, were taken within a few weeks in early 2012. The photos in Mexico and Cuba were taken some years earlier.

This book has benefited from significant contributions from several people. Rip Kirby provided the oblique aerials of the northeastern Gulf Coast and access to Eglin Air Force Base. The staff of the State of Mississippi Coastal Management Program took the photos of the Mississippi sea islands. Ervin Otvos provided these photographs and also contributed significantly to the text for the chapters on Alabama and Mississippi. J. W. Tunnell provided most of the photos of the coast of Mexico. Fabio Moretzsohn did most of the drawings. The manuscript benefited greatly from the scrutiny of Ervin Otvos and Ping Wang. The staff of Texas A&M University Press was very helpful and cooperative, especially senior editor Shannon Davies and illustration editor Kevin Grossman.

PART

GENERAL CHARACTERISTICS AND DYNAMICS OF BEACHES

Part I discusses the various elements of beaches, how they are formed, and how they behave. Beaches have been favorite places to visit for recreation for centuries, they are important to the fisheries industry, and they have been significant during international conflicts. Because of the importance of beaches, coastal scientists have been studying these environments all over the world. The size, morphology, behavior, and the level of wave energy that impacts beaches and adjacent coasts vary widely.

When we visit the beach for whatever reason—children playing, teenagers body surfing, fishermen wading in the surf zone, or families having a picnic—it is helpful to understand how things work in this dynamic and energetic environment. We need to be aware of certain things to help us enjoy the beach more and to remain safe while doing so; some of these are associated with waves and currents, and others with animals that might cause harm.

1

Coastal Processes

WHEN we hear the word *beach*, the first thing that comes to mind is sand; the next is probably waves. Actually there are multiple processes that impact beaches and control their existence and appearance (figure 1.1). It is appropriate to begin with the most fundamental of these coastal processes: the weather. Then it is important to consider how the waves, which are a result of the weather, impact the beach. These waves also generate currents that are a major element of beach dynamics. Storms, especially hurricanes, are a significant factor in Gulf of Mexico beaches. A process that is always present but is not weather related is the ebb and flow of tides, but tides do not play a major role in Gulf Coast beaches.

Weather

The Gulf Coast is positioned in the latitudes that range from about 18° to 30° north of the equator. This range of latitudes experiences a fairly wide variation in weather patterns. As the seasons change, so do the weather patterns. During the summer the Gulf is within the Trade Winds belt, with the prevailing direction from the southeast. This is the time when tropical storms can impact this coast. In the winter the westerlies prevail as weather systems are moving from the northwest to the southeast. The changes from one pattern to another influence the way beaches respond to the wind and the waves produced by it.

Westerlies

In the midlatitudes, the weather typically moves from west to east. High-pressure systems tend to come from the higher latitudes and collide with low-

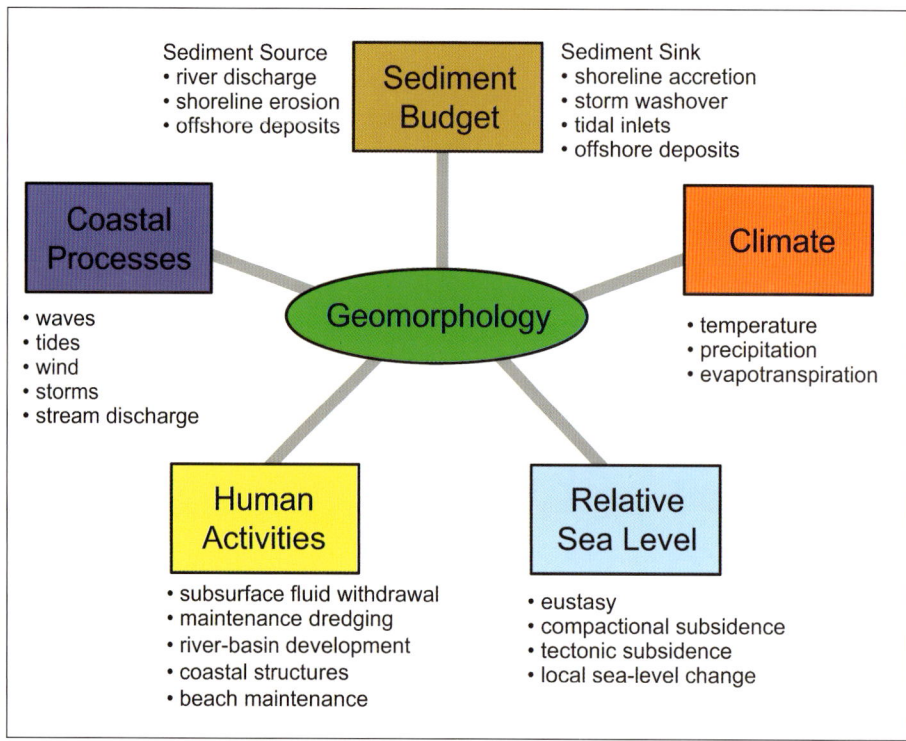

Figure 1.1. Various processes and other variables that constitute the beach system.

pressure areas, causing precipitation. It should be noted that wind tends to flow from high pressure toward low pressure. As these systems move across the mainland of the United States, they dominate the weather. For much of the year, this type of weather does not influence the Gulf Coast, but in the winter these patterns move into more southern latitudes. For example, the Florida peninsular coast experiences prevailing wind from the southeast from about mid-March to mid-October. As the seasons change, the sun moves to the southern hemisphere and the frontal systems of westerly weather impact this peninsula.

The typical situation occurs when a *cold front* (high pressure) comes from the northwest and passes across the Texas coast (figure 1.2a). As the front approaches, the wind is blowing onto the coast from the southeast. Just behind the front the wind is from the northwest and is typically strong. Lower temperatures and dryer air are associated with such frontal passages. On the Texas coast intense frontal passages are commonly called "blue northers" because of the cold north wind associated with them.

These frontal systems move across the Gulf of Mexico toward the Florida peninsula. There is some loss of intensity due to the warming effect of the Gulf

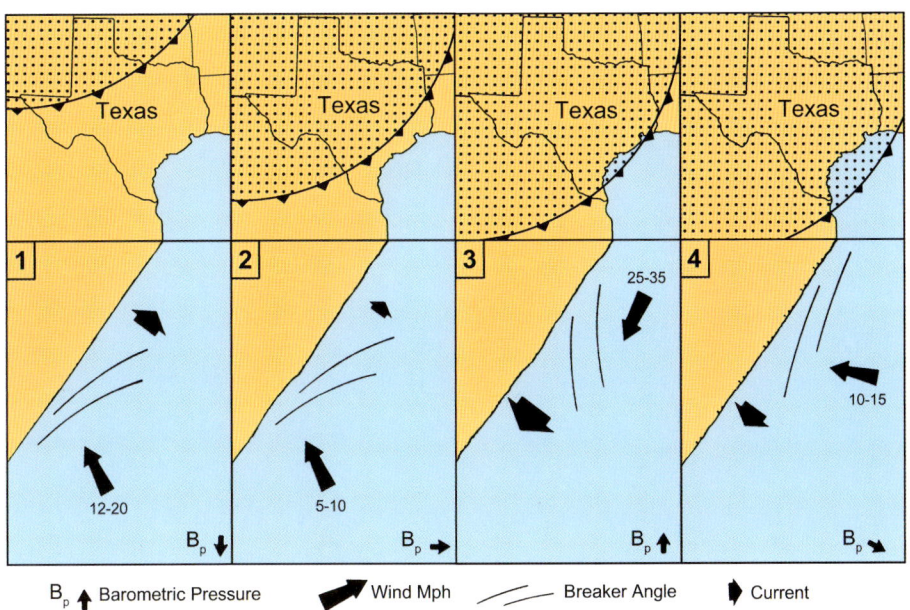

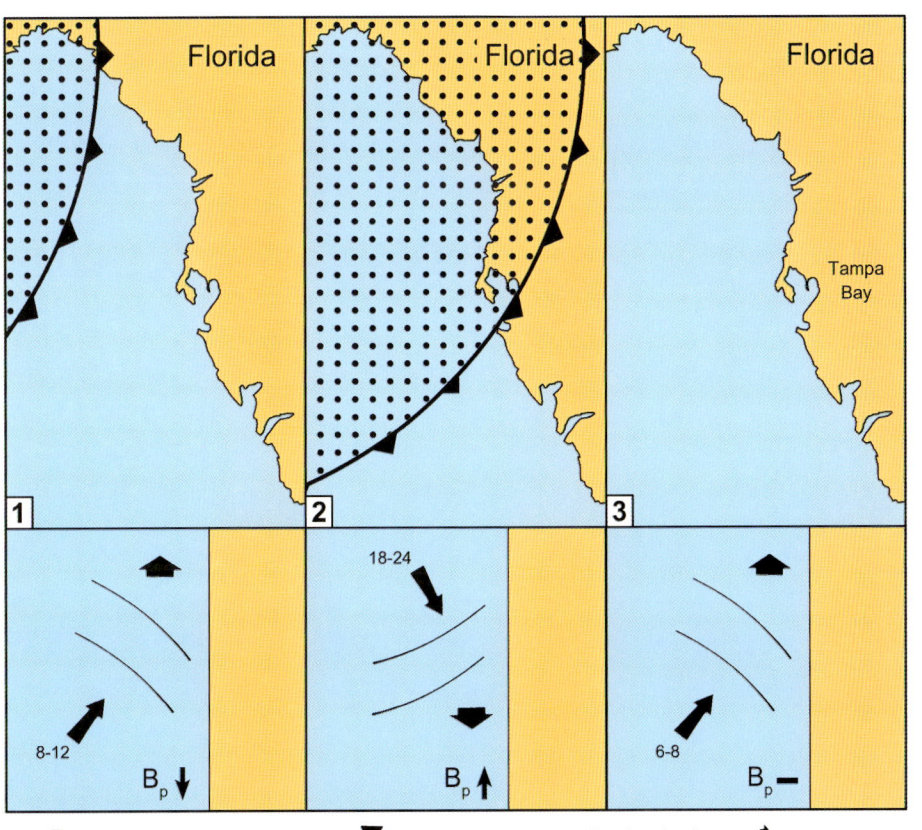

Figure 1.2. Frontal system as it passes (a) offshore of Texas and moves (b) onshore on the Florida coast;

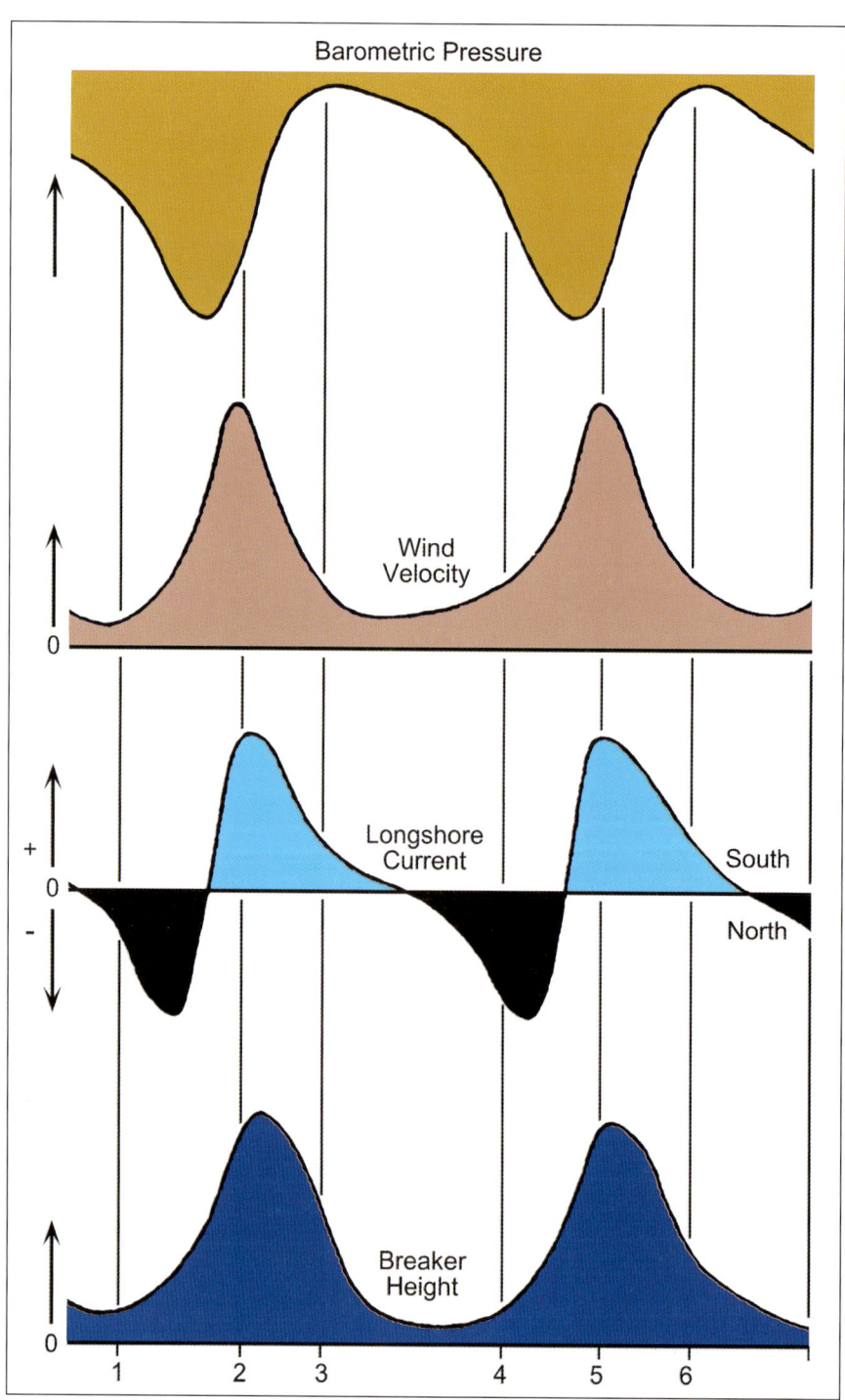

(c) general simulation model for coastal processes. From R. A. Davis and W. T. Fox, "Process-Response Patterns in Beach and Nearshore Sedimentation: I. Mustang Island, Texas," Journal of Sedimentary Research 45 (1975): 852–65.

water. As the front approaches the Florida coast, the wind has a southerly component. There is a rapid change to the northwest with increased speed as the front passes (figure 1.2b). Under some circumstances these fronts pass across the Florida peninsula and may move northerly along the Atlantic Coast to become a nor'easter, the strong storms of the winter in the New England area.

This pattern continues for several months, with the intensity of each frontal system increasing as the winter weather gets colder. Such frontal passages tend to dominate the wave and current patterns along the Gulf, especially on the Texas and Florida coasts (figure 1.2c).

Tropical Storms and Hurricanes

During the warm months, intense storms that develop in the tropical eastern Atlantic can impact the Gulf Coast. These storms begin as so-called tropical waves off the west coast of Africa. As the system gains intensity, it becomes a circulating low-pressure system. Because of its position in the low latitudes, it moves westward as part of the Trade Winds system. Circulation in these weather systems is in an anticlockwise direction. As the barometric pressure lowers, the wind speed increases, eventually reaching tropical storm level (39 mph) and in some cases eventually reaches hurricane status (75 mph).

The tropical storm/hurricane moves west and north as it approaches North America. As it passes the Caribbean Sea and the Bahamas, such a storm will commonly move in one of two paths: up to the north along the Atlantic Coast or into the Gulf of Mexico. Those storms that enter the Gulf can have a range of tracks (figure 1.3): some move westward to Mexico, some swing up onto the northern Gulf Coast, and rarely they move onto the west coast of the Florida peninsula.

These severe storms have a major influence on the coast, especially the beach environment. The speed of the wind and the size of the waves generated by this wind are dependent on the intensity of the storm. Hurricanes are rated on a scale that ranges from category 1 to category 5 as the wind speed increases. The increase of wind causes a related increase in the *storm surge*, the increase in water level produced by the friction of the wind over the water surface (figure 1.4). Such a phenomenon produces dangerous flooding in coastal areas during these storms. Storm surge at the coast is also related to the adjacent continental shelf. A shallow and wide shelf, such as on the Florida peninsula, will cause water to build up and increase the storm surge. A more narrow and steep shelf, such as the one adjacent to the Florida Panhandle, will have a much lower surge given the same wind conditions.

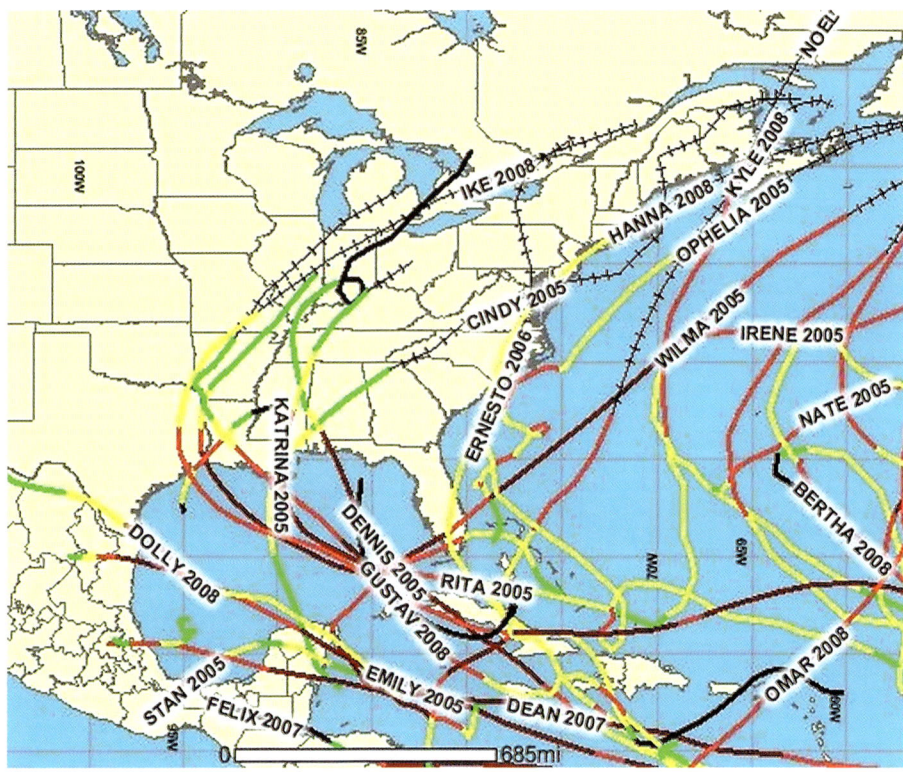

Figure 1.3. Tracks of hurricanes into the Gulf of Mexico since 2000. Courtesy NOAA.

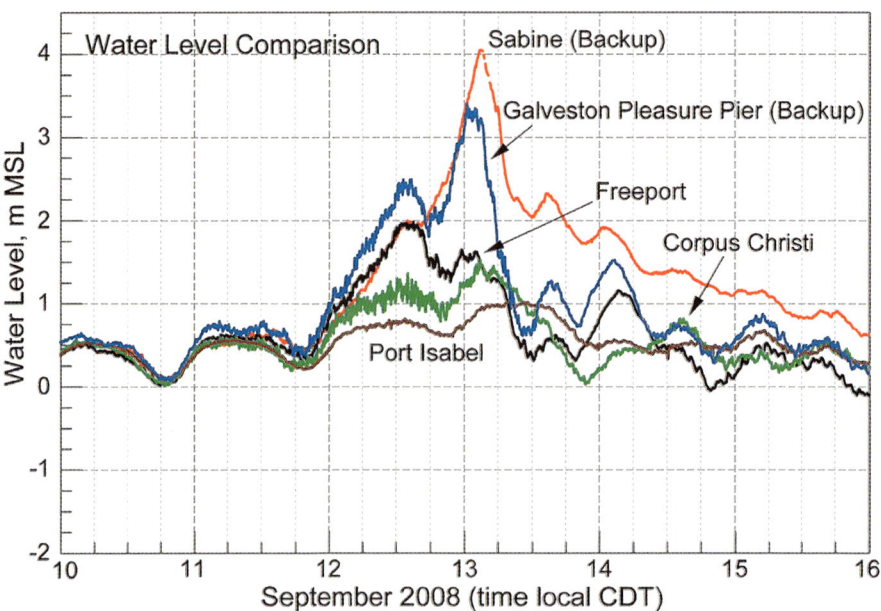

Figure 1.4. Plot of increase in water level at various locations caused by the passing of Hurricane Ike in 2008. Courtesy NOAA.

Because these intense storms have a cyclonic (anticlockwise) circulation, there is also offshore wind as the storm passes. Depending on the specific location of the coast relative to the storm system, there will be onshore wind producing a large storm surge as well as offshore wind that could diminish the storm surge.

Waves

The friction between the wind and water produces waves that are a regular disturbance of the water surface. Waves may occur at any location where wind blows over the water, but they are especially important along the coast. The beach is largely a result of wave action that may either erode the beach or build one. The size of waves in open water is the combined result of the speed and duration of the wind and the *fetch*, the distance over which the wind blows. In some basins there is a theoretical limit to the size of the waves because of the size of that basin. These fetch-limited basins include the Gulf of Mexico and the Mediterranean Sea. Large waves may develop in the Gulf, but there is a limit to how big they may be. Waves that are directly under the influence of the wind are called *seas*, and waves that have moved beyond the direct influence of the wind are called *swell*.

There is a wide range in the size of the waves as measured by the *period*, the time that it takes for a complete wave length to pass a point in space. In the Gulf of Mexico the *wave period* commonly ranges from about 3 seconds up to 6 or 7 seconds under normal conditions. Storms can generate waves that are significantly longer—up to 10 seconds in the Gulf. Wave period may exceed 20 seconds in the Pacific Ocean.

The motion of the water in a wave is essentially circular; that is, the wave form moves progressively, but the water does not. If it did, then we would have huge piles of water at the coast. This circular path of water within the wave at the surface has a diameter that is essentially equivalent to the height of the wave. The diameter decreases with depth until there is no motion at a depth of about one-half the wave length (figure 1.5). The *wave length* is the distance from crest to crest or trough to trough. A scuba diver likes to be at a depth equivalent to at least half the wave length. Less deep than that will cause the diver to move in a circular path as the water does and perhaps cause seasickness.

When waves move into shallow water and approach the shoreline, they are influenced by the bottom surface. As soon as the wave reaches a water depth equivalent to half the wave length, the bottom interferes with the wave motion

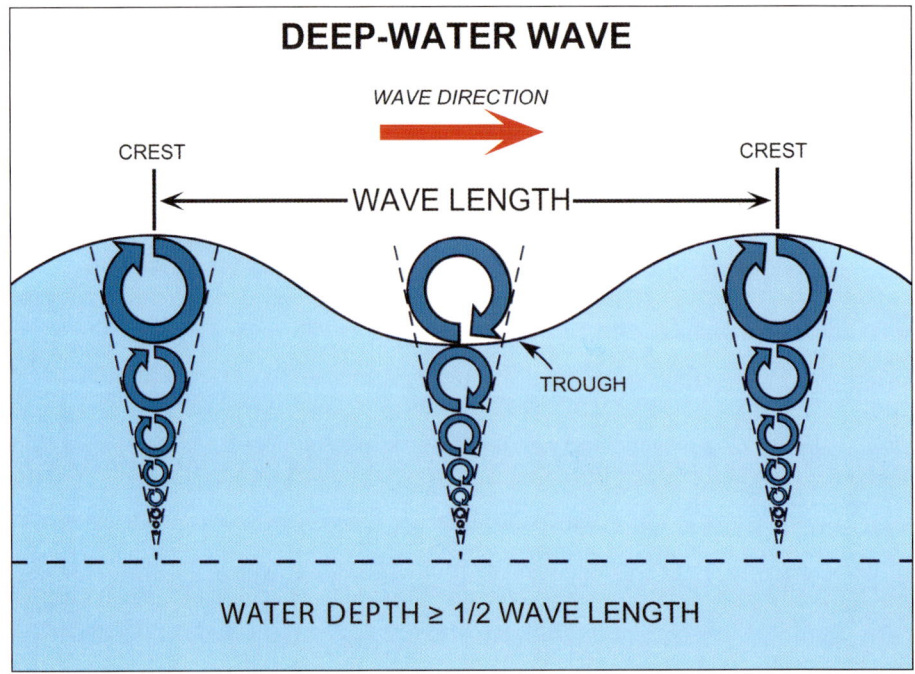

Figure 1.5. Parts of a wave and the water motion associated with waves.

and the circular motion becomes elliptical (figure 1.6). This causes the waves to compress in length and increase in height. As this process continues, the waves eventually become unstable and break. Breaking waves are very important in beach processes because they cause large amounts of sediment to become suspended and to move. There are three primary types of breaking waves: spilling, plunging, and surging (figure 1.7). Spilling waves break relatively gently and slowly, like spilling water from a glass. These breakers generally develop from waves that are directly under the influence of the wind. It is common for there to be multiple rows of spilling waves as they break over sandbars in the nearshore. Plunging waves break violently with a large "crash" and typically form from swell waves. Surging waves move up the foreshore as the last breaking wave on the beach.

Because wave breaking is the result of instability caused by shallow depths interfering with wave motion and propagation, there is an effect of water depth as the wave enters shallow water and approaches the beach. Along most coasts, especially in the Gulf of Mexico, sandbars are present parallel to the shoreline. There are commonly two or three sandbars, but the number varies. It is typical for waves to break over the bars and then re-form because all of

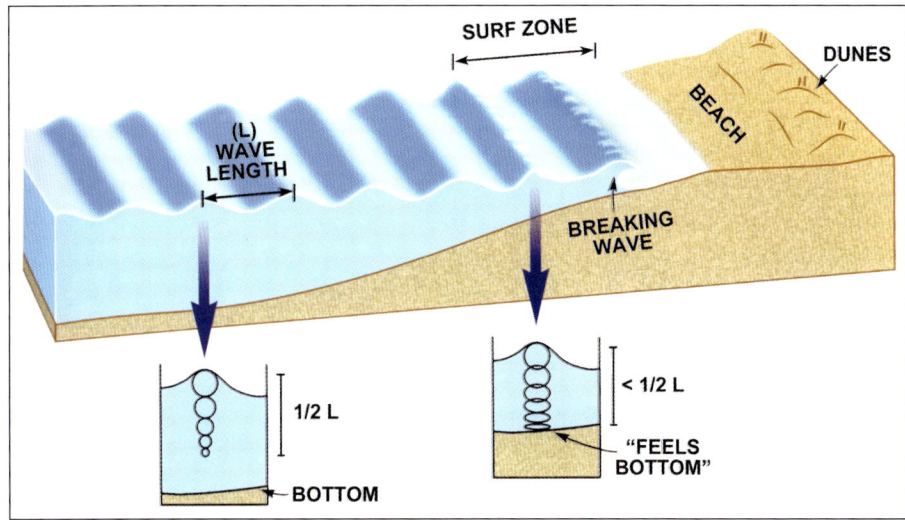

Figure 1.6. Diagram showing how the ocean bottom interferes with wave motion.

Figure 1.7. (a) Spilling breakers, (b) plunging breakers, and (c) surging breakers. Photo courtesy PDPhoto.org.

their energy is not lost during breaking (figure 1.8). As a consequence we can see multiple rows of breaking waves parallel to the shoreline (figure 1.9). The band of breaking waves is the *surf zone*.

Wave Refraction and Longshore Currents

There is another very important aspect of the waves moving into shallow water that has a major impact on the beaches—*wave refraction*, which is the "bending" of the wave crest as it moves into shallow water at an angle. It is quite rare for waves to approach the shoreline exactly parallel to it. Most commonly, waves approach at an acute angle to the coast based on the direction of the wind that generates these waves. As the wave enters shallow water and slows, eventually to break, it does so at different times and positions along the wave crest. This causes the wave to refract or bend (figures 1.10 and 1.11). This phenomenon causes water to be transported along the surf zone in the direction into which the angle of approach opens. These are called *longshore currents* and may transport considerable sediment. Their speed is dependent on the size and speed of the waves, along with the angle of approach.

Long-period swell waves may refract in relatively deep water because of their wave length. If this refraction is complete, the waves will approach almost parallel to the shoreline. Under these circumstances, there would be an insignificant longshore current and sediment transport.

The speed of longshore currents is generally highest in the troughs just landward of the sandbars. Under strong wind the speed may exceed a meter per second. The combination of the energy expended by breaking waves and the longshore currents can cause sediment, virtually all sand but with some shell material, to be carried along the shoreline in the surf zone. The rates of sediment movement in this zone can be quite high and may have a major influence on the dynamics of the adjacent beach. Various obstructions, some anthropogenic and some natural, can interrupt these currents and the sediment they transport.

▼ Figure 1.8. Oblique photo showing waves breaking over a single longshore bar.

▼▼ Figure 1.9. Waves breaking over longshore sandbars along the Texas coast. Photo by James F. Aber, Emporia State University.

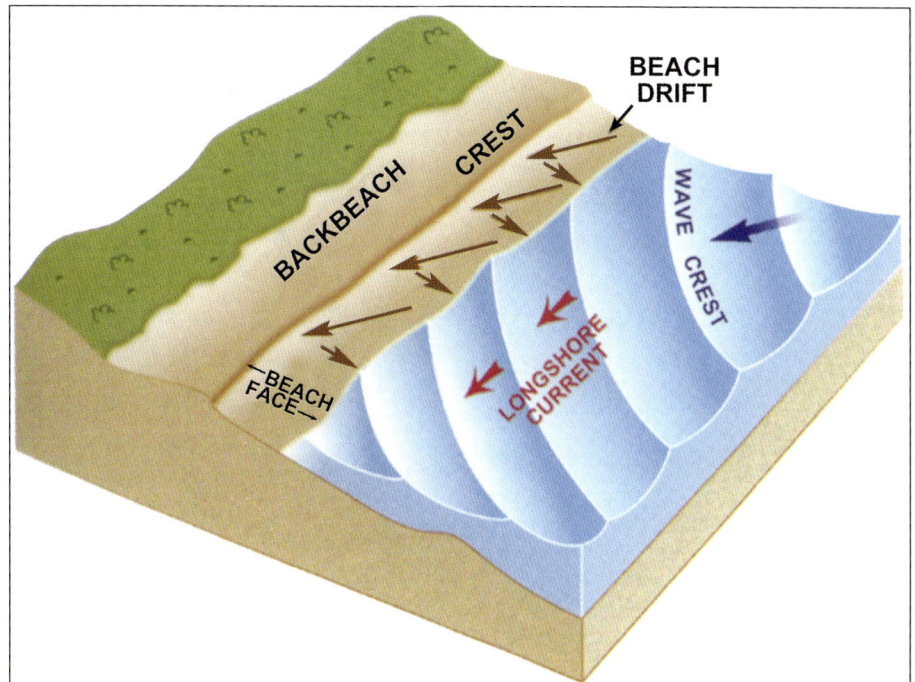

Figure 1.10. Diagram showing how wave crests refract (bend) as they enter shallow water and are slowed in their progression.

Figure 1.11. Aerial view of waves refracting as they pass through the nearshore and surf zone.

Rip Currents

Even though the basic water motion in waves is circular, friction from wind and wave breaking does cause some landward motion so that water builds up slightly at the shore. This is called *setup* and produces a modest instability in the water level at the shore. Because of this unstable and slightly elevated level, the water must return seaward. Under most circumstances it does so as *undertow*, where water is simply transported along the bottom. This is mostly associated with fairly steep beaches. A more common condition occurs where the seaward return flow is focused in a narrow concentration called a *rip current*.

The rip current may be directed through a low area, or *saddle*, in the crest of the longshore sandbar (figure 1.12), or it may be the result of converging currents that move along the shore. It is this phenomenon that causes problems for beachgoers and swimmers. Rip currents may be strong enough to carry swimmers seaward to depths beyond their ability to stand. The standard solution to the problem of being caught in a rip current is to swim parallel to the shore. These systems are narrow, and after a few strokes one should be free of their influence. Most beaches now have warning signs, and some have good information on recognition of rip currents. Generally, they can be recognized by actually seeing the water moving seaward (figure 1.13), or the waves are commonly lower over the rip because the seaward-directed water stifles them somewhat.

Rip currents only extend across the longshore bars in the surf zone, directed by the saddles in the bars. Beyond that there is really nothing that confines them, and the moving water spreads out. Sometimes it is possible to see this phenomenon because of the sediment that is in suspension. It is common for strong rip systems to carry sediment seaward, but the volume is not significant in the scheme of the beach's *sediment budget*.

Tides

Tides are the regular and predictable change in water level due to the gravitational attraction between the sun, moon, and earth. The relationship is proportional to the masses of these celestial bodies and inversely related to the distance between them. In other words, because the sun is so far away, about 93 million miles (150,000,000 km), its influence on the earth is relatively low even though it is tremendously large. The moon is quite small but is only 239,000 miles (385,000 km) away. As a result, its influence on the earth is about twice that of the sun's. The mass attraction between the earth and the moon is the main reason that we experience tides along the Gulf Coast. It is also the reason that tidal cycles follow the lunar cycle.

COASTAL PROCESSES   15

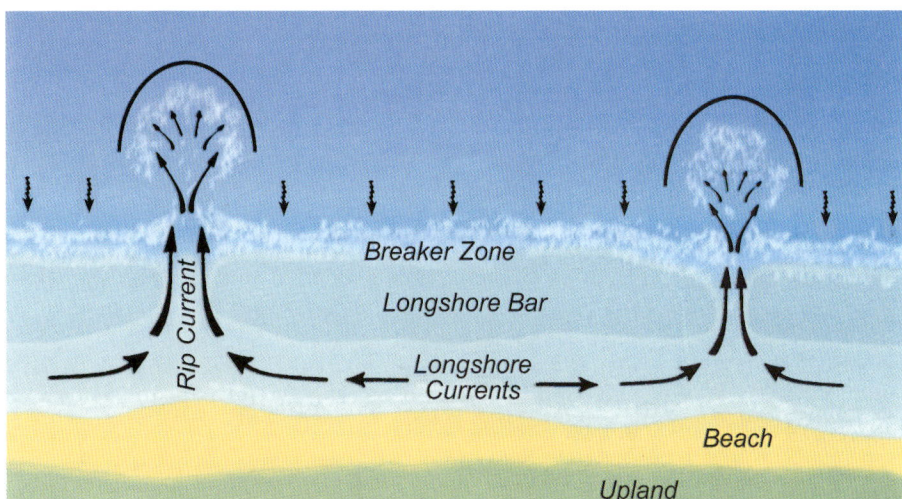

Figure 1.12.
◀ (a) The paths of water in a rip current system superimposed on the nearshore bathymetry, and
▼ b) large rip channels on the eastern coast of Australia.

The attraction between the earth and moon causes a bulge in the water of the earth's oceans. As the earth rotates, the bulge moves toward the shoreline twice each day: once on the side of the earth toward the moon, and the other on the opposite side (figure 1.14). As the lunar cycle changes over a lunar month, the amount of distortion in this bulge also changes. The maximum distortion occurs during the new moon and the full moon, a period of essentially two weeks. Under these circumstances the lunar and solar tides are superimposed.

Figure 1.13. Rip current where water is moving offshore between breaking waves.

The minimum distortion occurs during the first quarter and third quarter of the monthly cycle when lunar and solar tides are at right angles. The tides during the maximum are called *spring tides*, and the minimum tides are *neap tides*.

In general, the change in water level during each tidal cycle is small in the Gulf of Mexico—less than 1 m on nearly the entire coast. Even so, the difference between spring tides and neap tides can be significant. *Semi-diurnal* tidal cycles occur twice daily, and *diurnal* tidal cycles occur only once a day. The Gulf of Mexico experiences mixed tides during a tidal month (figure 1.15); that is, the tides are diurnal on some days and semi-diurnal on others.

Along some coasts the tides are very large. In the United States the panhandle of Alaska has by far the largest tides, up to 12 m. In the lower 48 states, tides along the Oregon, Washington, and Maine coasts reach nearly 4 m. In some of these places the beach is essentially a wide tidal flat (figure 1.16). On the Gulf Coast things are quite different. The tides are quite small, and the intertidal zone is very narrow. It is generally easy to recognize the high tidal level when one goes to the beach. It commonly is marked by seaweed, shells, or other materials. The tidal stage does not, however, have a major impact on how we use the beach. About the only thing we need to know is where to put our beach chair or blanket so that the tide does not come up and cover it.

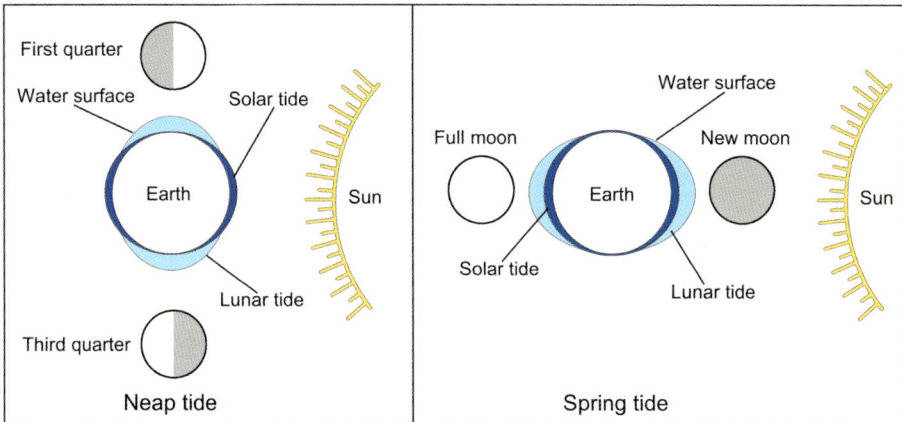

Figure 1.14. Relationships between the earth, sun, and moon that produce neap and spring tides.

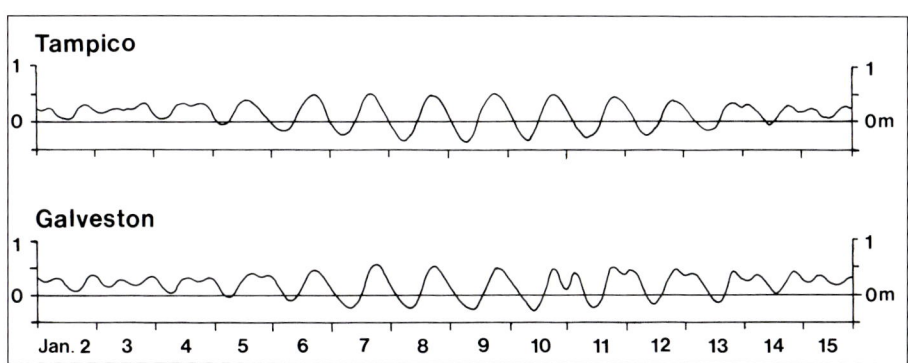

Figure 1.15. Tidal record at Tampico, Mexico, and Galveston, Texas, showing the typical mixed tidal patterns of the Gulf of Mexico.

Sea Level

Although a very slow process, sea-level change is an important factor in the beach environment. This topic has been prominent recently in the various media due to an apparent increase in the rate of sea-level rise related to the warming of the earth's climate. Actually, both the rise and fall of sea level have an impact on the beach. As sea level falls, the beach tends to become wider and may be accumulating significant sediment. This is not the situation at the present time on the Gulf Coast, where erosional conditions dominate, but it is happening in the high latitudes of the northern hemisphere due to glacial rebound. As the glaciers have melted, the crust rises as it adjusts, and the result is a relative lowering in sea level. This adjustment takes place as a part of *eustasy*. The northern part of Scandinavia is experiencing this phenomenon now. In

North America there are numerous beach ridges representing former shorelines along the margin of Hudson Bay in Canada (figure 1.17).

Most of the world is currently experiencing a rise in sea level, and the Gulf Coast is no exception. The glaciers have been melting for about the past 20,000 years, and we have a good record of the rise in sea level that has resulted from that and related phenomena such as warming of the ocean. The initial rate of sea-level rise was rapid, about 1–2 cm per year. About 6500–7000 years ago the rate of rise slowed to about 2–3 mm per year, and then around 3000 years ago sea level reached near its present position (figure 1.18).

During the past 20 or so years there has been an increase in the rate of

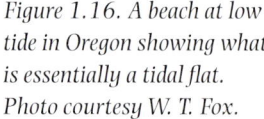

Figure 1.16. A beach at low tide in Oregon showing what is essentially a tidal flat. Photo courtesy W. T. Fox.

COASTAL PROCESSES   19

Figure 1.17. Aerial photo showing numerous beach ridges formed along Hudson Bay, Canada, while sea level was falling. Photo courtesy A. Hequette.

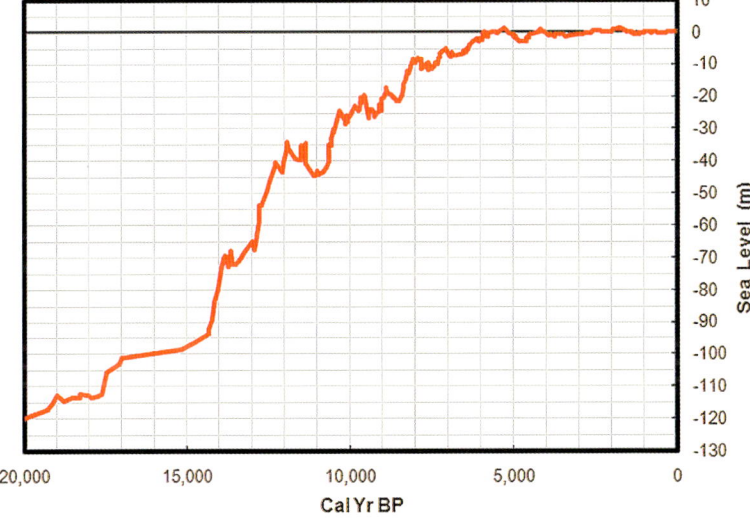

Figure 1.18. Sea-level curve showing the pattern of sea-level rise since the glaciers began to melt about 20,000 years ago. Courtesy Florida Geological Survey.

sea-level rise, from about 1.5 mm to more than 3 mm per year (figure 1.19). This increase has caused concern to all who live and recreate along the coast. Such an increase, if it persists over centuries, will cause significant erosion to the beach environment. The increase in sea level will bring the shoreline higher on the beach and expose it to wave attack, which can cause erosion. It will also decrease the width of the beach and may move the shoreline landward.

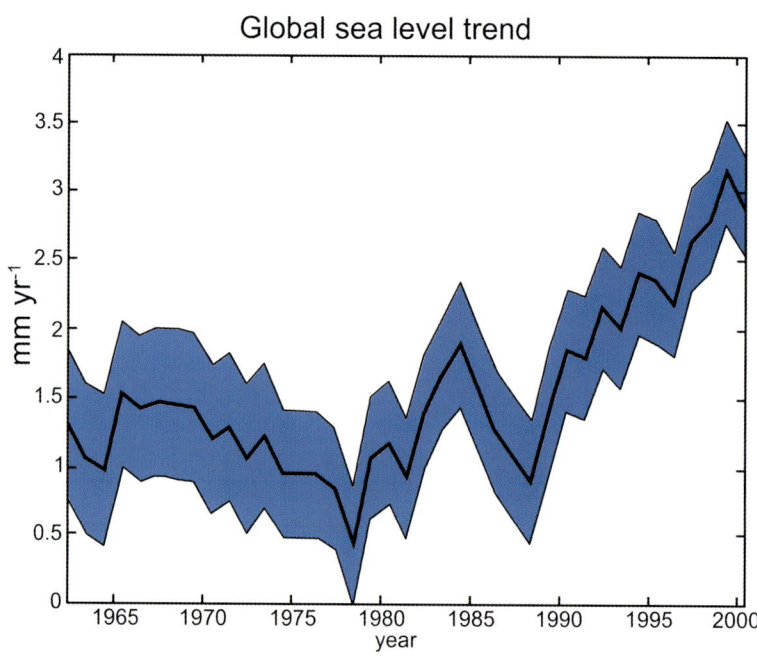

Figure 1.19. Increase in sea-level rise during the end of the twentieth century. Diagram modified from M. A. Merrifield, S. T. Merrifield, and G. T. Mitchum, "An Anomalous Recent Acceleration of Global Sea Level Rise," Journal of Climate 22 (2009): 772–81.

SUGGESTED READING

Davidson-Arnott, R. 2010. *Introduction to Coastal Processes and Geomorphology*. Cambridge: Cambridge University Press.
Davis, R. A. 1994. *The Evolving Coast*. New York: Scientific American Publishing.
Davis, R. A., and D. M. FitzGerald. 2004. *Beaches and Coasts*. Oxford: Blackwell Publishing.
Hayes, M. O., and J. Michel. 2008. *A Coast for All Seasons*. Columbia, SC: Pandion Press.

2

Beach Geomorphology and Barrier Island Morphodynamics

BEACHES are one of the most dynamic of all surface environments on the earth. Changes can take place in literally seconds. Major changes are commonly the result of severe storms in only a day or so. This chapter discusses the nature of the beach, its *morphology*, the process-response systems that cause changes, and the way the beach interacts with the shallow nearshore environment in the seaward direction and the adjacent dunes in the landward direction.

Beach Morphology

Looking at a profile across the beach and its adjacent environments, we find a complex of subenvironments. The beach and nearshore zones as discussed here start seaward with the outer longshore sandbar. This nearshore zone extends to the low-tide line where the beach begins. The landward limit of the beach is located where the morphology changes to a dune environment or, if there is human development, perhaps a seawall. Beginning in the nearshore, subtidal environment are multiple longshore sandbars and intervening troughs (figure 2.1). These longshore bars are typically parallel to the shoreline and are wave formed. There are also places where the shallow water steepens and breaking waves commonly occur (figure 2.2). The number of these longshore bars typically ranges from one to three depending on the availability of sediment and the slope of the nearshore. A gradual slope tends to have three, and a relatively steep slope will commonly have only one (figure 2.3). The number of longshore bars at a given location tends to be the same through time. Waves generally break only over the innermost bar except under storm or

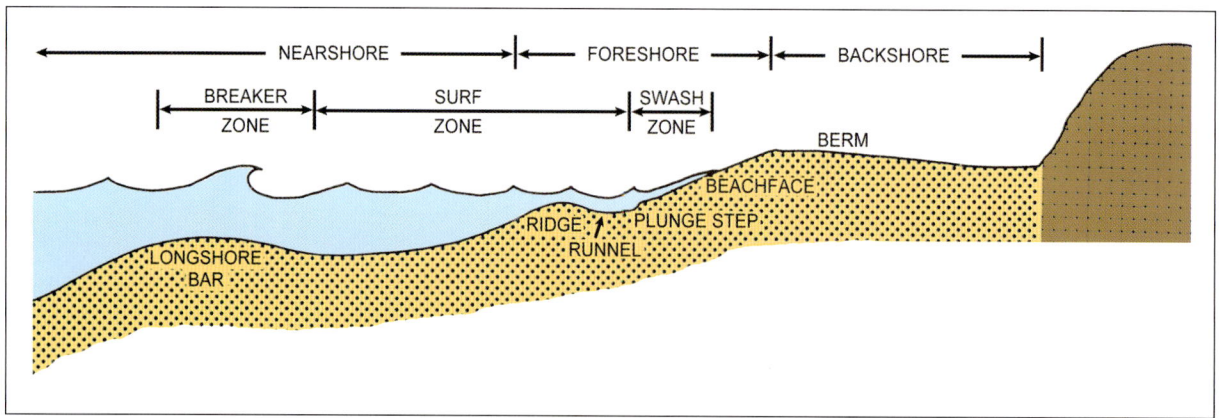

Figure 2.1. Profile across the beach and adjacent nearshore environment showing subenvironments.

Figure 2.2. Waves breaking over multiple longshore bars on Mustang Island, Texas.

Figure 2.3. Waves breaking over a single longshore sandbar.

high-wave conditions. The less energetic troughs between the bars are where surf fishermen cast their bait in hopes of catching something.

These longshore bars tend to be continuous for significant distances along the shoreline, but in some places the inner bar will merge with the shoreline. The crest of the bars is not at exactly the same level throughout its extent. There may be *saddles* where rip currents tend to develop (figure 2.4). In some situations these pathways might be actual channels if the rips are strong enough to cause sand to erode. Figure 2.5 shows an extreme example of this condition.

Although the longshore bars just described are the typical sandbar

Figure 2.4. Multiple rip current channels along the coast.

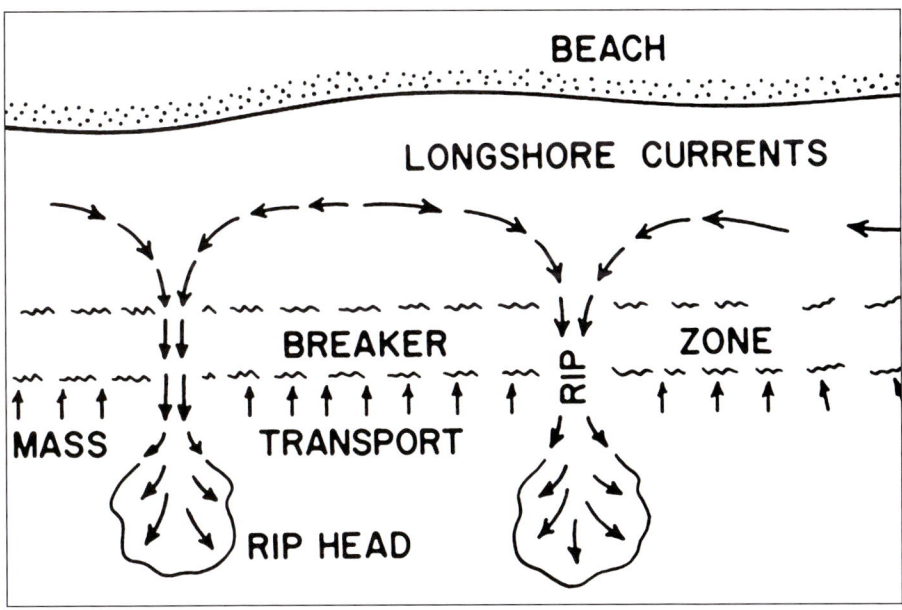

Figure 2.5. Diagram showing the circulation in the rip current system. Arrows represent the direction of water motion.

configuration and have the typical shore-parallel position, other morphologies of nearshore sandbars include attached bars, oblique bars, and transverse bars (figure 2.6). *Attached bars* occur where part of the bar is attached to one of the seaward bulges or protuberances on the shoreline. This condition tends to be temporary and quite localized. *Oblique bars* are attached to one of the protuberances and typically terminate in the surf zone. Rip currents tend to form in the channels that are present between adjacent oblique bars (figure 2.7). *Transverse bars* are simply extensions of the beach protuberance into subtidal waters. All four types of nearshore bars have been reported from the Florida Panhandle, but they also occur elsewhere.

Ridge and Runnel

Longshore bars are subtidal. Landward of these are ephemeral, intertidal bars called *ridges*. The *runnel* is a shallow trough that separates the ridge from the landward beach (see figure 2.1). These intertidal features represent the natural post-storm beach-recovery profile. The sand is removed from the beach and accumulates in the adjacent intertidal zone (figure 2.8). Ridges and runnels may also develop as waves accumulate linear sand bodies in shallow subtidal water and then transport the sand landward (figure 2.9). This sediment transport is caused by wave-generated currents carrying sediment across the ridge so that it migrates much like a bedform. The breaking, surging waves create such currents, and depending on the tidal range, it takes from about two

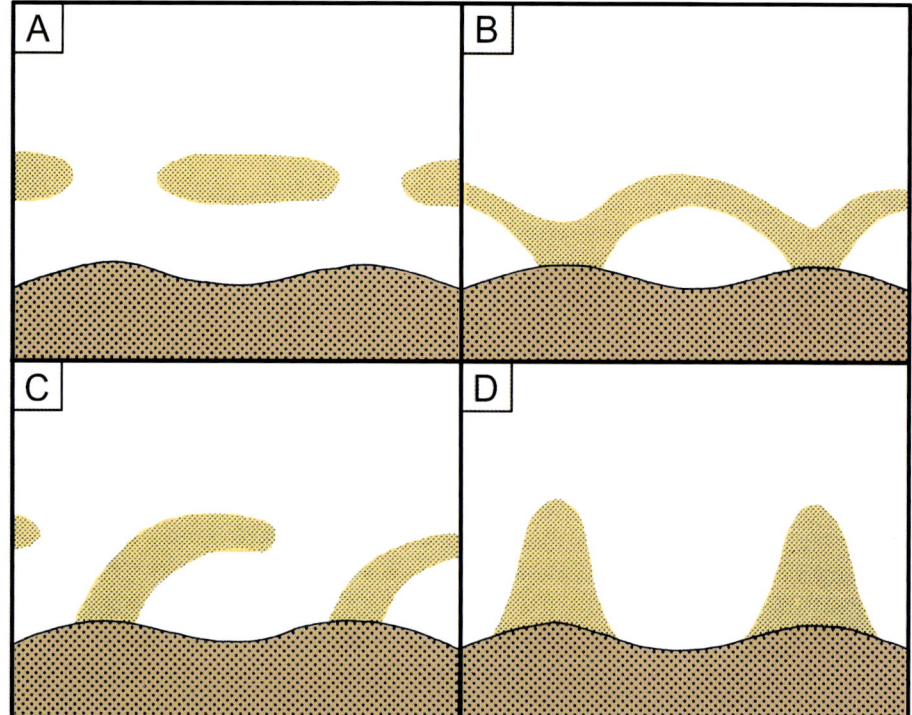

Figure 2.6. The four types of nearshore sandbars: (a) shore-parallel, (b) attached, (c) oblique, and (d) transverse.

Figure 2.7. Intertidal oblique bars where rip currents develop on the Oregon coast. Courtesy W. T. Fox.

Figure 2.8. Large ridge and runnel on the Florida coast after Hurricane Elena in the fall of 1985.

Figure 2.9. Typical appearance of the ridge and runnel along the Padre Island coast of Texas.

Figure 2.10. Migration of a ridge onto the upper intertidal zone, where it becomes part of the foreshore beach.

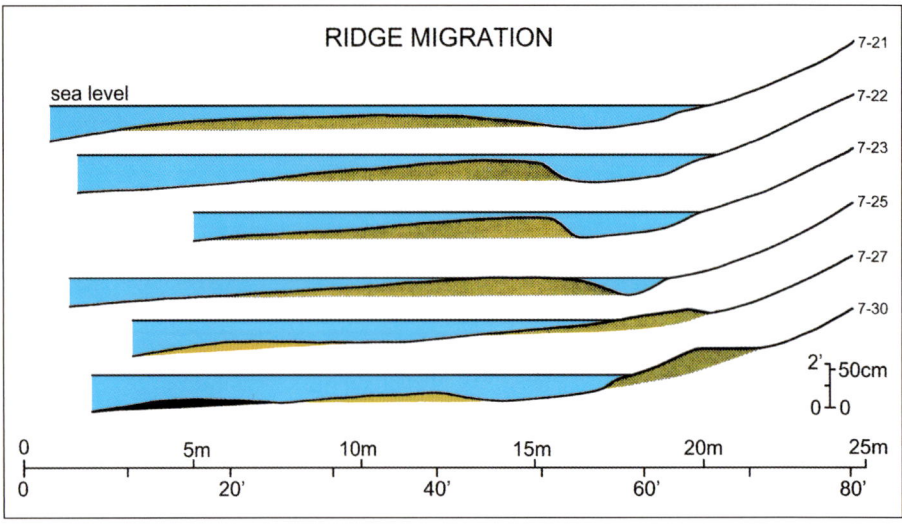

weeks to more than a month for this ridge to migrate onto the upper intertidal beach (figure 2.10). Sometimes smaller multiple ridges may form if abundant sediment is available.

Foreshore

Once the ridge "welds" onto the beach proper, it is subjected to the back-and-forth movement of water in the swash zone. The ridge and runnel is considered here as part of the intertidal beach. The *swash zone* is the next landward

Figure 2.11. Swash on the foreshore portion of the beach.

component of the beach, where the last energy of the waves is expended with a continual uprush and backwash of water (see figure 2.1).

The relatively steep part of the beach on which the swash zone resides is the *foreshore* (figure 2.11). It slopes seaward at varying angles. The foreshore is probably the most dynamic of the beach subenvironments because it is continually being subjected to wave action. The foreshore of the beach responds directly to the wave energy at the time. It may be wide and gently sloping as is common during times of beach construction when wave energy is generally low. It is generally steeper as wave energy increases and the beach becomes erosional. Grain size also influences the foreshore. Coarse-grained beaches are steep relative to fine-grained beaches.

Backshore

The most landward part of the beach is the *backshore*. This "dry beach" is essentially horizontal and is wide on a prograding beach (figure 2.12). Its width depends on the availability of sediment and the wave energy. After storms the dry beach may be gone due to erosion, and then the entire beach is the foreshore and is wet. It is also narrow and relatively steep.

Figure 2.12. Excellent example of a dry backbeach on the Florida coast.

Figure 2.13. Coppice mounds on Padre Island, Texas. These represent the first stage in the development of coastal dunes on the backbeach.

Opportunistic vegetation starts on the backshore, and *coppice mounds*, the small sand accumulations that are precursors to dunes, will develop (figure 2.13). Because of their size they are vulnerable to storms, but they also may form again in only months. The foredunes generally are fairly stable and

Figure 2.14. Numerous foredune linear trends (dark lines) on Matagorda Island. Courtesy Google Earth.

hold their position for many years. They commonly form well-developed linear patterns (figure 2.14).

Shoreline

The configuration of the shoreline does show some patterns. The largest scale of these exhibits *rhythmic topography*, a sinuous shoreline pattern that resembles a low-amplitude sine wave (figure 2.15). The origin is not well understood but is undoubtedly related to the wave climate. This shoreline configuration can last for several months but will be destroyed by a severe storm.

One of the most common shoreline features, *beach cusps*, is much smaller in scale than rhythmic topography. The cusps consist of a seaward part called the *horn* and an intervening bay (figure 2.16). The wave length of the cusps ranges from about 6 m up to 20 m. There are two kinds of cusps: one where the cusps and horns are sand throughout (figure 2.17a) and the other where the horn is dominated by gravel (figure 2.17b). Wave action is involved in some way in each type, but the one dominated by sand throughout may be formed mostly by erosional processes. In contrast, the gravel horns are depositional features. They seem to be present in low-energy conditions and are shorter lived than the sand cusps.

Figure 2.15. The regular and sinuous shoreline that is called rhythmic topography. Each of these elements can range up to hundreds of meters in wave length. This example comes from the Florida Panhandle. Photo by US Geological Survey.

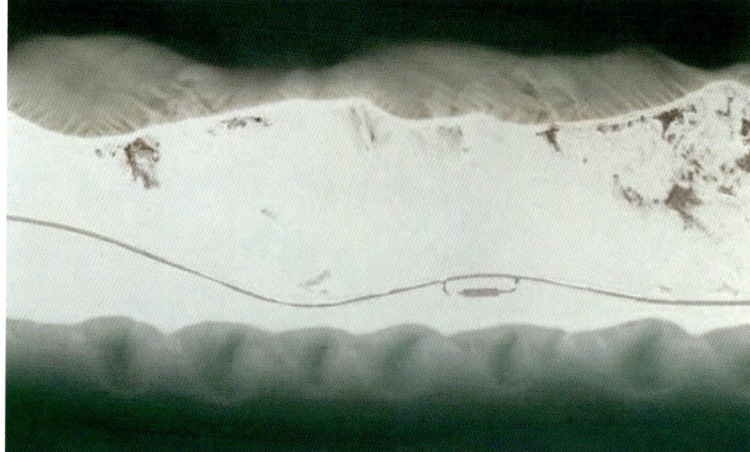

Figure 2.16. The nature of beach cusps with horns and bays. The scale ranges from a few meters to tens of meters.

▼ *Figure 2.17. Two types of beach cusps: (a) one is essentially sand throughout, and (b) the other has horns of gravel.*

Storm Impact on Beaches

Because the beach is a fragile, exposed, wave-dominated coastal environment, elevated wave-energy and storm surge as a result of increased onshore wind during storm activity can cause major changes to the beach. The high energy of storm waves, which are larger and steeper and move faster than usual, can easily remove sediment from the beach. Storm conditions also affect longshore bars in the adjacent nearshore. Sand tends to be removed from the bars, they are somewhat flattened, and they move offshore in the range of tens of meters. We are still learning about how the wave/nearshore bottom systems behave during storms. The high-energy conditions in the surf zone make it difficult to monitor the system during a storm. As the storm subsides, the beach environment tends to return toward pre-storm conditions.

The beach shows marked changes after storms. There is significant removal of sediment from the foreshore and backbeach (figure 2.18). The post-storm profile shows a generally uniform and relatively steep profile. There is commonly a ridge and runnel present. As time passes, this ridge can migrate landward and merge with the storm beach. Over time and in the absence of another storm, the beach profile returns to a configuration similar to that before the storm. If, however, the storm is quite energetic and carries beach sand far offshore, the previous beach profile may not return completely to its pre-storm size and configuration.

The storm effects just described result in beach sediment moving off the beach in a seaward direction. It is also possible for the opposite situation to occur—beach sand being carried landward, sometimes in huge quantities. The general term given to this process is *overwash*. The resulting sediment accumulations are called *washover fans*, which may coalesce to form a *washover apron*. Rather severe storms are required for this process to occur. The storm surge and accompanying combined flow currents carry both water and sand across the beach and may cover nearly all of a barrier island (figure 2.19), or the washover may extend into a back-barrier bay (figure 2.20). Under these circumstances most of the sediment removed from the beach during a storm is carried landward, not into the surf zone.

Sediment is in limited supply in some locations along the Gulf Coast beaches. One such place is at Sargent Beach on the Texas coast where limited sediment has been carried onshore, exposing older marsh mud in the surf zone (figure 2.21). There is no barrier here, and considerable erosion has taken place with many homes destroyed.

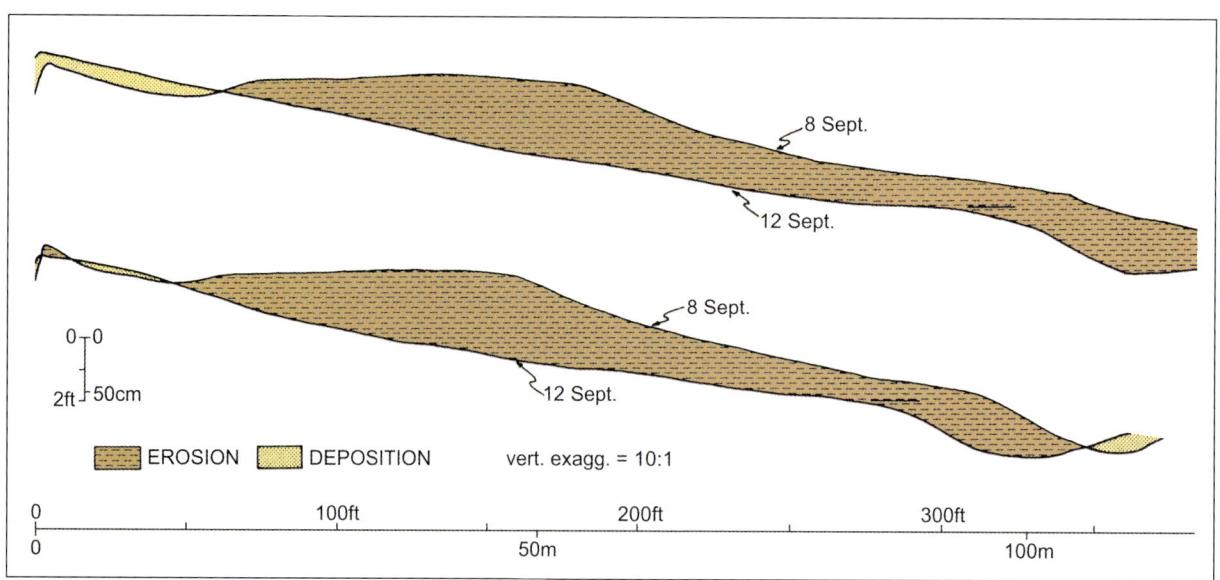

▲ *Figure 2.18. General beach profile showing pre- and post-storm configurations for Hurricane Fern on the central Texas coast in September 1971. From R. A. Davis, "Beach Changes on the Central Texas Coast Associated with Hurricane Fern, September, 1971," Bulletin of Marine Science 16 (1972): 89–98.*

▶ *Figure 2.19. Aerial view of a washover/blowover apron on Santa Rosa Island, Florida.*

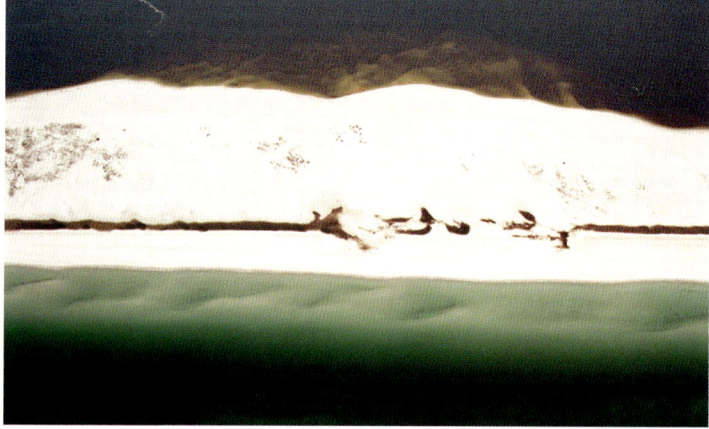

▶ *Figure 2.20. Aerial view of washover fans that extend into the bay over Dauphin Island, Alabama. Photo courtesy D. Nummedal.*

Figure 2.21. Erosional conditions at Sargent Beach, Texas, where (a) erosion has exposed small cliffs of marsh mud and (b) a small amount of sand has been transported landward.

Rocky Coast

Parts of the western and northeastern coasts of the United States consist of extensive rock exposures with cliffs, sea caves, wave-cut terraces, and cliffs (figure 2.22). The US Gulf of Mexico coast is nearly without any rocky areas. An exception is the Point of Rocks area on Siesta Key, Florida (figure 2.23), a section of coast about 300 m long composed exclusively of *beachrock*. This material was deposited as very shelly sand in the beach environment and was lithified (changed to rock) in place in a very short period of time. The cement in this particular location has been carbon dated at 4000–2600 years before present.

This is the age of the lithification, which might have taken only a decade or so to precipitate. The combination of near-tropical climate and abundant calcium carbonate in the form of shells will lithify the sediment rapidly.

This beachrock system extends from the present beach out for about 100 m into the Gulf. The nature of the cement and its age suggest that the material that forms the beachrock was associated with an older barrier island than the present Siesta Key, which is only 3000 years old. This older barrier was eroded,

Figure 2.22. Rocky coast along the northern California coast.

▼*Figure 2.23. (a) Aerial view of short coastal reach where beachrock stabilized the coast on Siesta Key, Florida. (b) Close-up of beachrock at the Siesta Key site.*

Figure 2.24. Rocky volcanic coast of Veracruz area in Mexico. Photo courtesy J. W. Tunnell.

but the lithified beach material remained and is now associated with the present barrier system.

There are also a few places in the Big Bend area of Florida where Miocene limestone crops out along the coast. These are quite small and local, and they are covered by marsh sediment and plants. Such locations are not really what would be considered rocky coastal areas.

The coasts of Mexico and Cuba have a much wider range of shoreline types, including volcanic and other types of rocky shores. Some of these are very rugged and have high relief (figure 2.24). There may be an absence of beach-type material and also a place where beach material can accumulate and be stable. There are sections of the Campeche coast of Mexico where Tertiary limestone forms the shoreline. Beachrock is also present on the Yucatán Peninsula coast.

Barrier Island Systems

Now that beaches have been described and discussed, it is appropriate to consider how the beach environment is integrated into the bigger picture: that of the barrier island system. Virtually all Gulf of Mexico beaches are associated with barrier islands. The emphasis here is on the *morphodynamics*, or change with time, of the major geomorphic elements of this system: the barrier islands and the tidal inlets that separate them. In addition to the beach and nearshore zone, barrier islands have dunes and washover fans (figure 2.25). The washover fans may have wetlands and tidal flats on them, or they may be *supratidal*,

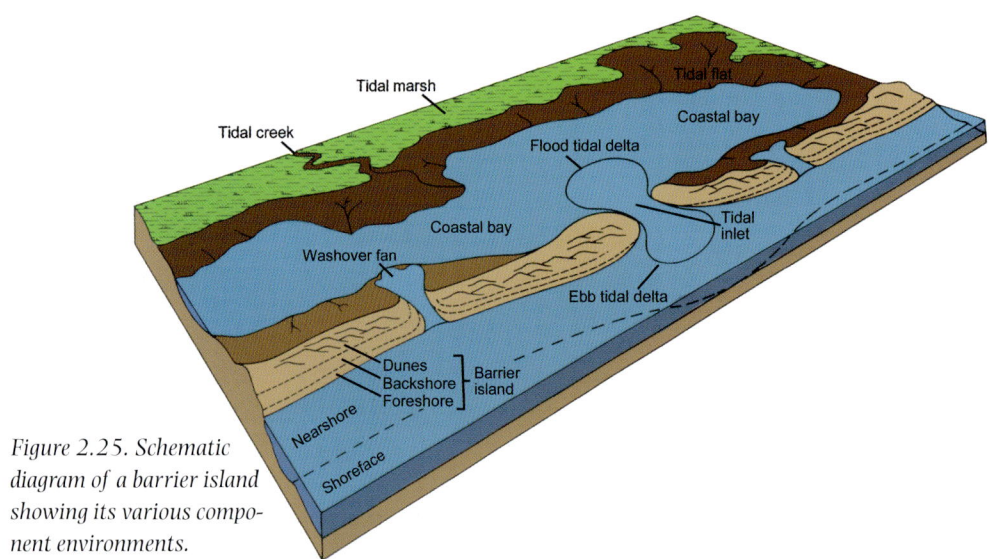

Figure 2.25. Schematic diagram of a barrier island showing its various component environments.

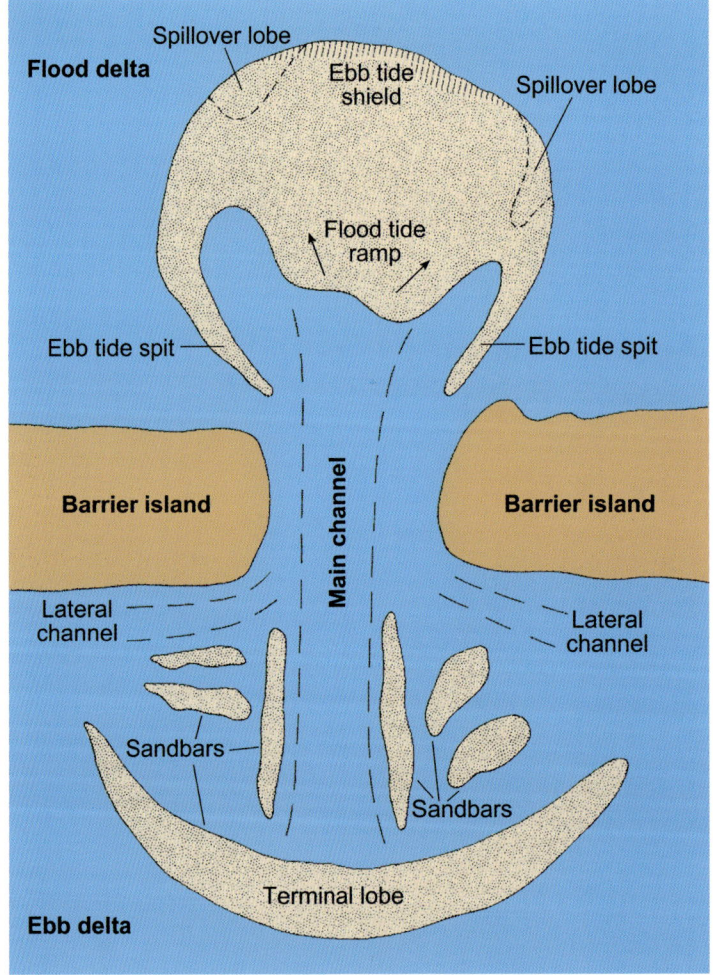

Figure 2.26. Diagram showing the three major elements of a tidal inlet along with numerous minor environments.

just above high tide. Tidal inlets have three primary geomorphic elements: main channel, flood-tidal delta, and ebb-tidal delta (figure 2.26).

Barrier Islands

There are various types of barrier islands depending on their shape, which is related to the combination of wave and tidal processes. The two primary types are *wave-dominated barriers* and *mixed-energy barriers*, with some variation within each. Those that are wave dominated tend to be long, up to several tens of kilometers, such as Padre Island on the Texas coast (figure 2.27). This is the result of wave processes that generate longshore currents that distribute sand along the coast. Wave-dominated barriers may be narrow or wide depending on their elevation and the availability of sand. Some accumulate much sand that becomes incorporated into dunes. As time passes, these dunes form ridges, and with considerable sand, several ridges may form (figure 2.28). Where wave-dominated barriers have less sand, only small dunes develop, which permits large storm waves to wash over the barrier and transport sand from the beach and surf zone to the landward side in the form of washover fans (figure 2.29). This condition can lead to fairly wide barrier islands. Some barriers have both of these conditions.

The other primary barrier shape is a "drumstick" barrier, a mixed-energy barrier island. The morphology of this barrier is dependent on a combination of wave and tidal processes. These barriers tend to be relatively short with one wide end and one that is narrow (figure 2.30). The development of such a shape is the result of sand being captured by wave-generated currents at one end of the island and therefore not being able to be transported down the barrier shoreline by longshore currents (figure 2.31). The narrow end of these drumstick barriers is susceptible to washover due to their low elevation and narrow width because of lack of sediment. These barriers are *progradational* on the wide end and transgressive, or landward moving, on the narrow end.

Tidal Inlets

Tidal inlets may also have a range of size and shapes depending primarily on tidal prism and longshore transport of sand. The *tidal prism* is the volume of water that passes through the inlet during each flood- and ebb-tidal cycle. This volume is dependent on the tidal range and on the area of the water body or bodies landward of the barrier(s) that is served by the inlet. In the Gulf of Mexico this volume may be quite small, only a few hundred thousand cubic meters of water or as much as tens of millions of cubic meters. On the Florida

Figure 2.27. (a) Padre Island, Texas, one of the longest barrier islands in the world. (b) Anclote Key, Florida, a good example of a wave-dominated barrier island.

(a)

(b)

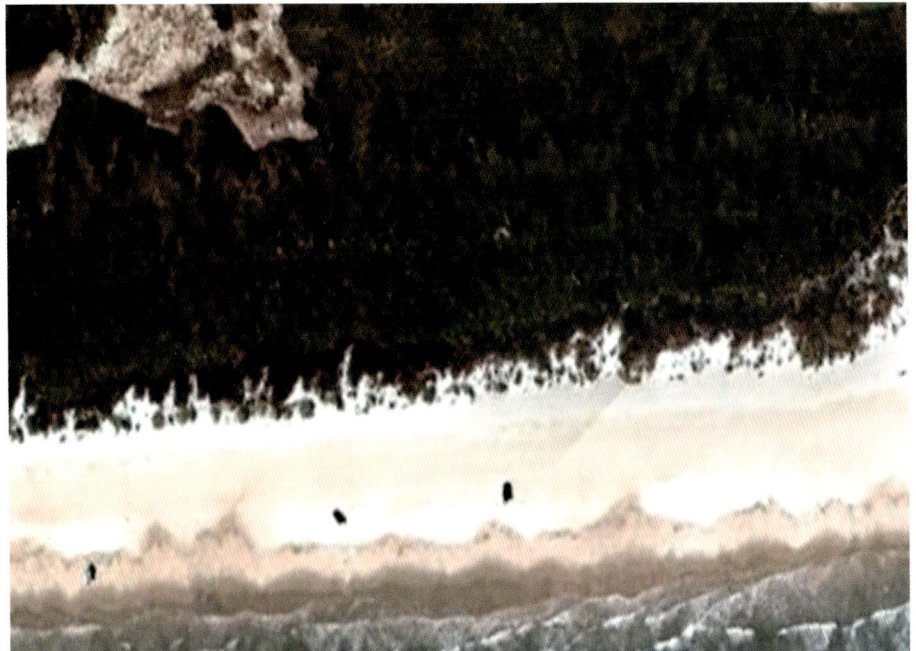

Figure 2.28. Multiple dune ridges can make a barrier wide through seaward progradation, such as seen here on Mustang Island, Texas. Courtesy Google Earth.

Figure 2.29. Storms and large waves can carry sand across the barrier and deposit it in the form of washover fans, such as here on the Panhandle of Florida.

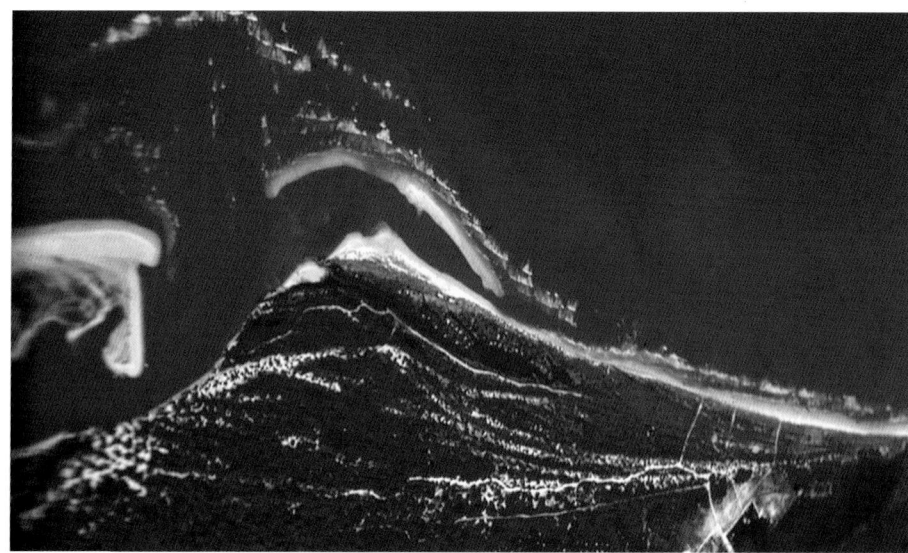

Figure 2.30. Caladesi Island, Florida, a drumstick or mixed-energy barrier island with a wide prograding end and a narrow end.

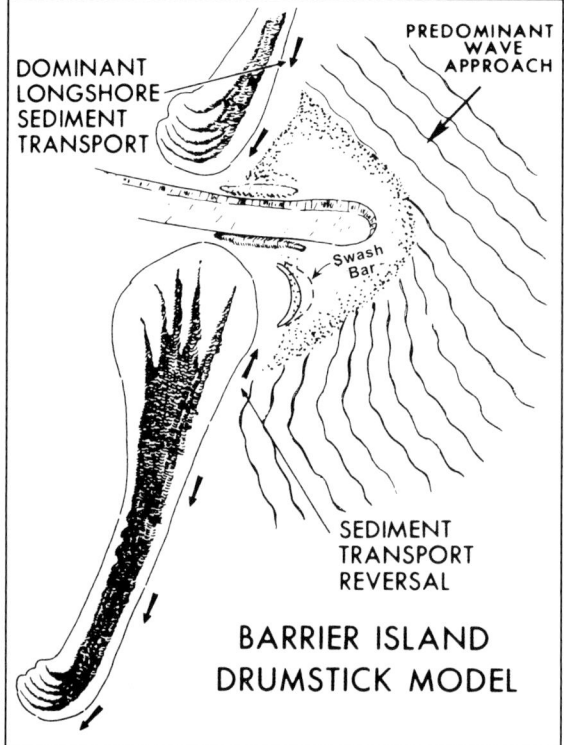

Figure 2.31. Wave and current patterns that form the drumstick shape to a barrier island. From M. O. Hayes, "General Morphology and Sediment Patterns of Tidal Inlets," Sedimentary Geology 26 (1975): 139–56.

GEOMORPHOLOGY AND MORPHODYNAMICS   41

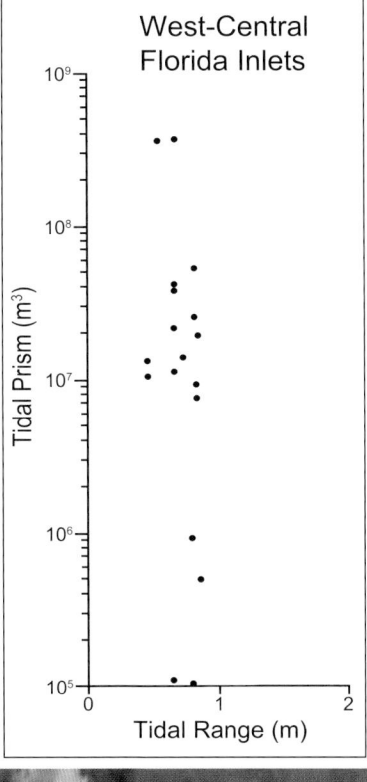

Figure 2.32. Plot of tidal range and tidal prism for inlets on the Gulf Coast of the Florida peninsula, showing the wide range in prism due to the wide range of bay area.

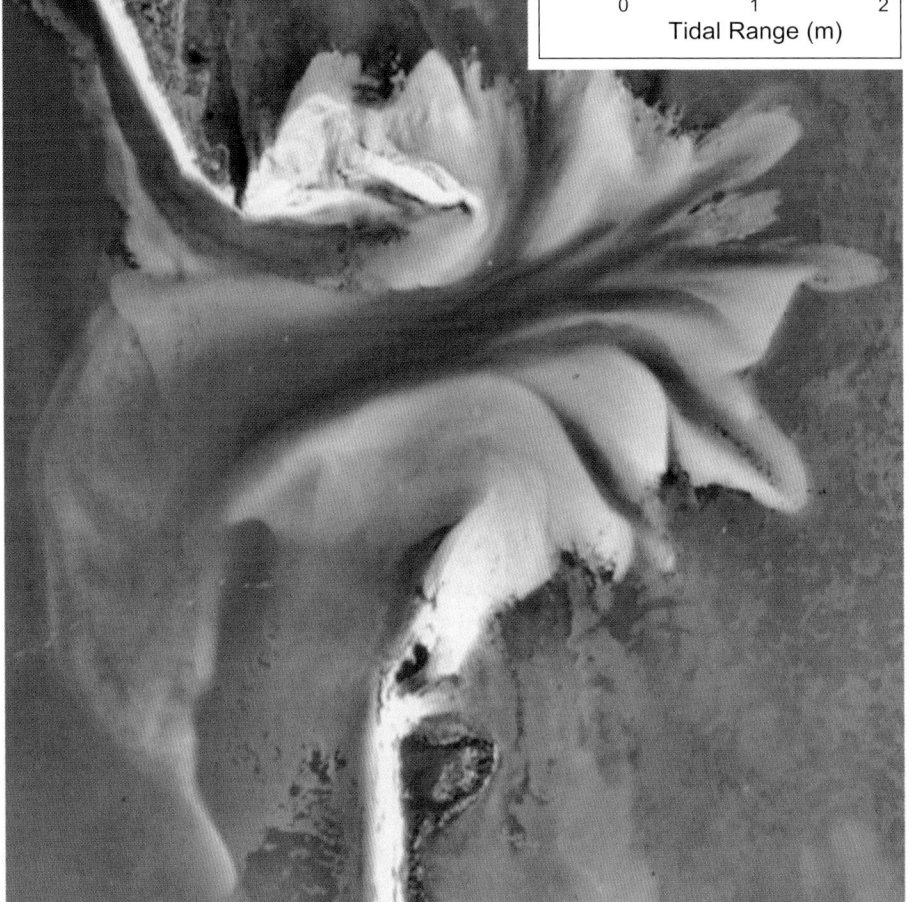

Figure 2.33. Aerial view of the fan-shaped flood delta at Hurricane Pass, Florida, as it appeared in 1951.

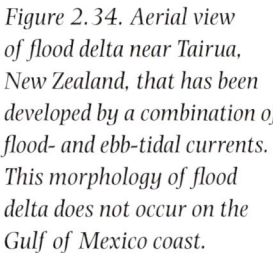

Figure 2.34. Aerial view of flood delta near Tairua, New Zealand, that has been developed by a combination of flood- and ebb-tidal currents. This morphology of flood delta does not occur on the Gulf of Mexico coast.

peninsula Gulf Coast the range of tidal prism covers four orders of magnitude (figure 2.32). The stability of the tidal inlet is dependent on the ability of the tidal currents, which are directly in proportion to the tidal prism, to keep the inlet open and in a constant position.

The tidal currents in the main channel of the inlet may transport huge amounts of sediment. Much of this sediment may be deposited at the landward and seaward ends of the inlet channel. The sediment body on the landward side is the *flood-tidal delta* and on the seaward end is the *ebb-tidal delta* (see figure 2.26). The flood delta takes on two different morphologies: one is fan-shaped, and the other looks somewhat like a horseshoe crab. The difference is primarily controlled by the tidal range. The fan-shaped flood delta occurs where tidal range is low, typically less than 1 m (figure 2.33). These are generally subtidal, and ebb-tidal currents flow over them without any significant influence.

The other shape of flood-delta bodies is limited to places where the tidal range is above 1.5 m. Under these conditions the ebbing tidal currents will mold the sediment so that there are ebb spits that look a bit like tails (figure 2.34). Between these tails is an ebb shield that prevents the ebbing currents from moving over it and causing them to be deflected to each side. This type of flood delta does not occur on the Gulf of Mexico coast.

Tidal inlets occur in three morphologies with one having two varieties: tide-dominated, mixed-energy, and wave-dominated (figure 2.35). The ebb-delta area determines which of the inlet types is present. *Tide-dominated inlets* have a large ebb delta that protrudes out into the Gulf (figure 2.36). These are

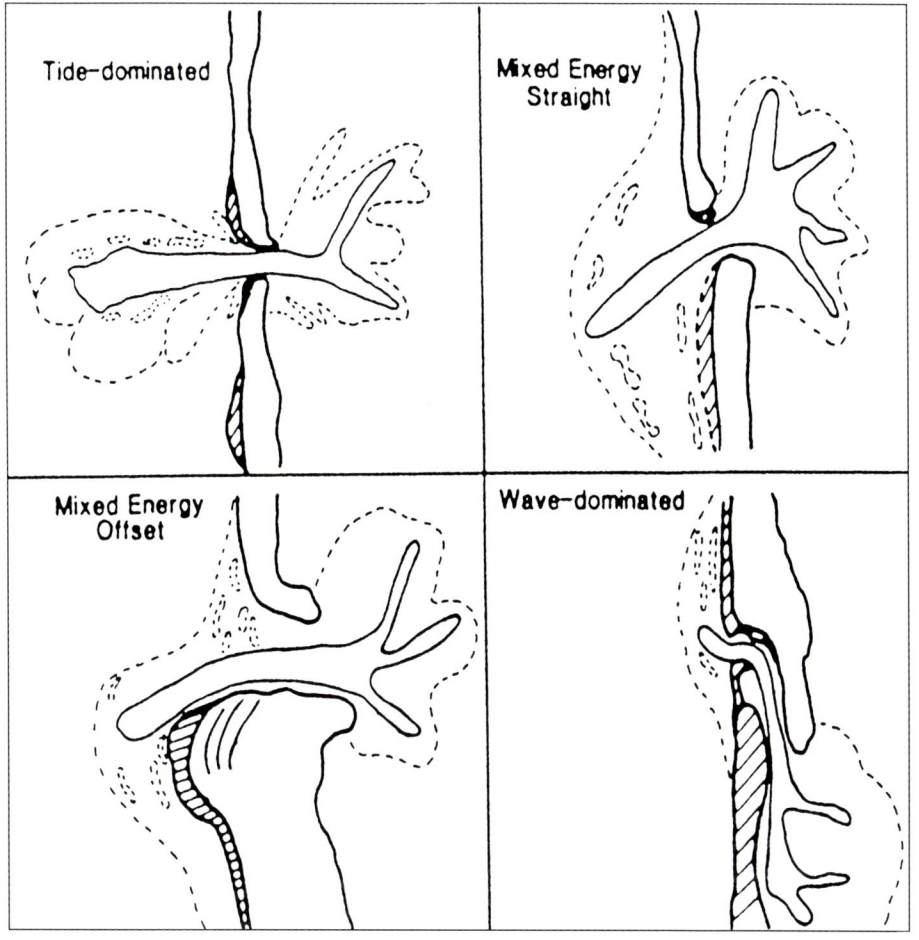

Figure 2.35. Classification of tidal-inlet morphology that applies well to the Gulf of Mexico coast. After J. C. Gibeaut and R. A. Davis, "Statistical Geomorphic Classification of Ebb-Tidal Deltas along the West-Central Florida Coast," special issue, Journal of Coastal Research *18 (1993): 165–84.*

Figure 2.36. Bunces Pass tidal inlet near the entrance to Tampa Bay, Florida, is a good example of a tide-dominated inlet. Note the large ebb-tidal delta.

typically deep and incised stable inlets characterized by a large tidal prism. *Mixed-energy inlets* have a substantial ebb delta but with a wave-modified terminal lobe. There are small marginal flood channels on each side along the adjacent barrier shoreline (figure 2.37). These inlets also tend to be stable, and they occur in two forms: one that has a straight shoreline on each side of the channel and another that has a significant offset between the adjacent barrier islands. The *wave-dominated inlets* are small and unstable, with a small or nonexistent ebb delta. They are commonly terminated by closure due to longshore transport of sediment (figure 2.38).

Barrier Island / Tidal Inlet Interactions

The wave and tidal processes that operate along the open coast and form the morphology of barrier islands and tidal inlets generate considerable interaction

◄ *Figure 2.37. Mixed-energy tidal inlets have marginal flood channels (arrows) and a main ebb channel that terminates in a sediment body called the terminal lobe. The example here is New Pass, Florida.*

► *Figure 2.38. Example of a wave-dominated tidal inlet at Midnight Pass, Florida, as it appeared in 1983. It closed a few months later and remains so.*

between these two major sedimentary environments and greatly influence barrier beaches. Sediment exchange between inlets and barriers may cause the adjacent beach to increase in width or, in extreme cases, be removed completely.

Inlets serve as a major interruption in the "river of sand" along the beach. They are fairly efficient sediment traps with much sand accumulating in both flood- and ebb-tidal deltas. Many natural inlets permit sand to bypass the inlet by moving across the ebb delta, especially in wave-dominated and mixed-energy inlets (figure 2.39). Tide-dominated inlets present a significant barrier to sand bypassing because they tend to block and trap sediment with their protruding ebb delta. Structures such as jetties are even greater barriers to longshore sediment transport. Some coastal researchers consider tidal inlets to be the biggest factor in beach erosion because of their restriction of longshore sediment transport.

Those tidal inlets that have a morphology that permits bypassing can provide considerable continuous "nourishment" to downdrift beaches. It is common to see an attachment where sand has moved across the terminal lobe of the inlet and onto the beach. This is a major factor in the development of drumstick barriers.

Figure 2.39. Oblique view of Big Sarasota Pass showing the sand body where sand is being transported to the next barrier island in the distance.

SUGGESTED READING

Bird, E. C. F. 2000. *Coastal Geomorphology: An Introduction.* New York: John Wiley and Sons.
Davidson-Arnott, R. 2010. *Introduction to Coastal Processes and Geomorphology.* Cambridge: Cambridge University Press.
Davis, R. A. 1994. *The Evolving Coast.* New York: Scientific American Publishing.
Davis, R . A., and D. M. FitzGerald. 2004. *Beaches and Coasts.* Oxford: Blackwell Publishing.ww

3

Beach Materials, Structures, and Sources

THE most common term associated with the beach is *sand*. True, most beaches are predominantly sand, but there are many other kinds of materials that can also be present in large amounts at some locations. In fact, the term *sand* denotes only grain size; it tells us nothing about the composition of the particles. Sand can be composed of a wide range of minerals. This chapter discusses the range of materials that constitute beaches: their textures, composition, and origins. This information will give us a much more comprehensive appreciation of the beach environment.

Beach Textures

Sand is a particle that is between 0.0625 mm and 2.00 mm, or about $1/16$ inch. This range of particle size is part of a comprehensive size classification called the Wentworth Grain Size Scale (table 3.1). Some of the terms for grain-size categories in this classification are quite recognizable, but they also have specific quantitative definitions. For example, the term *boulder* has a specific definition: any particle between 256 mm and 1048 mm or about 10 inches and larger. The terms *cobble*, *pebble*, *silt*, *clay*, and *mud* also have specific quantitative size ranges. Beaches can be composed of boulders, cobbles, or sand (figure 3.1).

Two grain-size terms warrant special explanation. One is *gravel*, which refers to all grain sizes larger than sand (>2.0 mm). Most of the time we think of pebble-sized gravel, but the term is not restricted to that category. The other commonly used term is *mud*, a mixture of silt and clay. It is important to remember that these two commonly used terms have specific definitions according to particle size (figure 3.2).

TABLE 3.1. WENTWORTH GRAIN SIZE SCALE

mm	ø units		Size class		
2048	−11		Very large		
1024	−10		Large		
512	−9		Medium	Boulders	
256	−8		Small		
128	−7		Large		
64	−6		Small	Cobbles	GRAVEL
32	−5		Very coarse		
16	−4		Coarse		
8	−3		Medium	Pebbles	
4	−2		Fine		
2	−1		Very fine	Granules	
1	0		Very coarse		
1/2	+1	500 μm	Coarse		
1/4	+2	250 μm	Medium	Sand	
1/8	+3	125 μm	Fine		
1/16	+4	62 μm	Very fine		
1/32	+5	31 μm	Very coarse		
1/64	+6	16 μm	Coarse	Silt	
1/128	+7	8 μm	Medium		MUD
1/256	+8	4 μm	Fine		
1/512	+9	2 μm	Very fine	Clay	

We typically speak about beach sediments by using their mean or average grain size. When a beach sediment is described as medium sand, the average grain size is in that range. This can be expressed in millimeters, but that approach is a little messy in terms of the very large range of numbers. In 1934 a professor at Northwestern University, William Krumbein, suggested using the $-\log_2$ as a better way to express grain size. That means every size category is doubled or half of the adjacent one. This approach provided grain size in single-digit numbers called ϕ units.

Another important textural aspect of beach sediments is their *sorting*. This is a quantitative expression of the uniformity of grain size at a particular location or in a specific sample of beach material. In a statistical sense, it is the

MATERIALS, STRUCTURES, AND SOURCES 49

Figure 3.1. (a) Boulder beach in the Bay of Fundy, Canada. (b) Cobble beach on the island of Tahiti. (c) A sand beach on the Florida coast.

standard deviation of grain-size measurements. People who study sediments conduct grain-size analyses and generally plot their distribution on graphs, either in bar graph form or as a linear curve (figure 3.3). If the distribution is narrow and the curve is narrow and steep, the sediment is well sorted(figure 3.4). When the distribution is wide, the sediment is poorly sorted.

Many beaches along the Gulf of Mexico are a combination of two types of materials: typically shell gravel and sand. This grain-size situation is called

Figure 3.2. ▶
Sargent Beach, Texas, which has a mixture of shell gravel, sand, and mud.

Figure 3.3. ▼
Distribution curve of beach sediment showing good sorting and poor sorting.

Figure 3.4. ▶▼
Close-up of well-sorted sand, which is typical of Gulf Coast beach sediment.

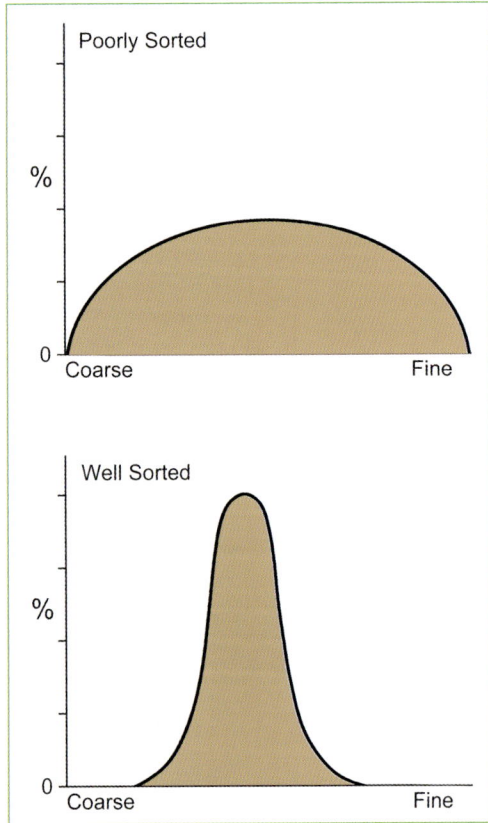

bimodal because each of the two types has a mode, or the most common grain size (figure 3.5). Each may be well sorted, but the sample as a whole is poorly sorted.

The grain size of the beach sediment has an influence on the profile of the beach. This is primarily true in the foreshore zone where wave swash moves up and down across this relatively steep part of the beach. The general relationship is that the coarser the beach sediment, the steeper the slope of the foreshore (figure 3.6).

MATERIALS, STRUCTURES, AND SOURCES   51

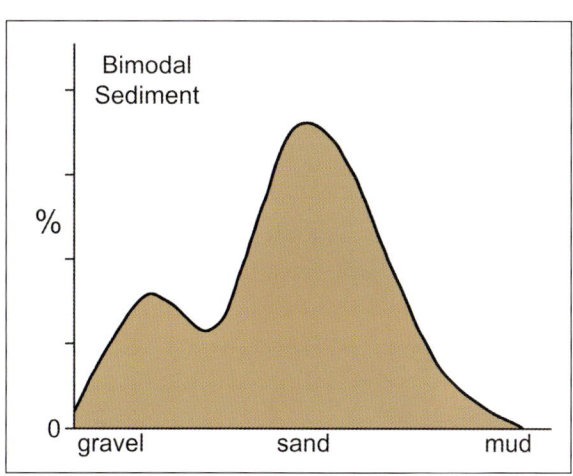

Figure 3.5. An example of a bimodal distribution plot for a beach sediment.

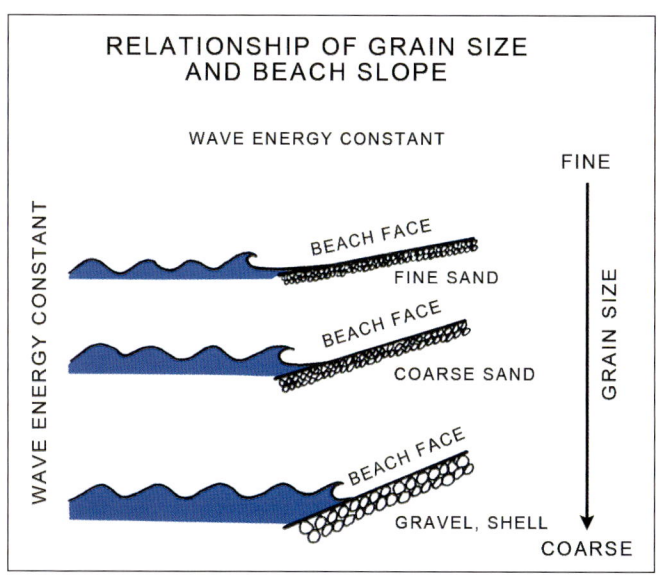

Figure 3.6. General relationship between beach sediment grain size and the slope of the beach foreshore.

(a)

(b)

Figure 3.7. (a) A steep, shelly foreshore and (b) a very gentle foreshore slope of a fine sand beach showing the relationships in figure 3.6. Photo a, courtesy J. W. Tunnell.

That relationship explains why we see shelly beaches with a steep foreshore (figure 3.7).

Thus far the discussion has considered bulk textural properties of sediments. We now consider the individual sediment grains with particular attention to their shape. Two attributes of grain shape are important in understanding beach and coastal sediment: sphericity and roundness. One is the three-dimensional nature of the grains. The most simple and practical way to

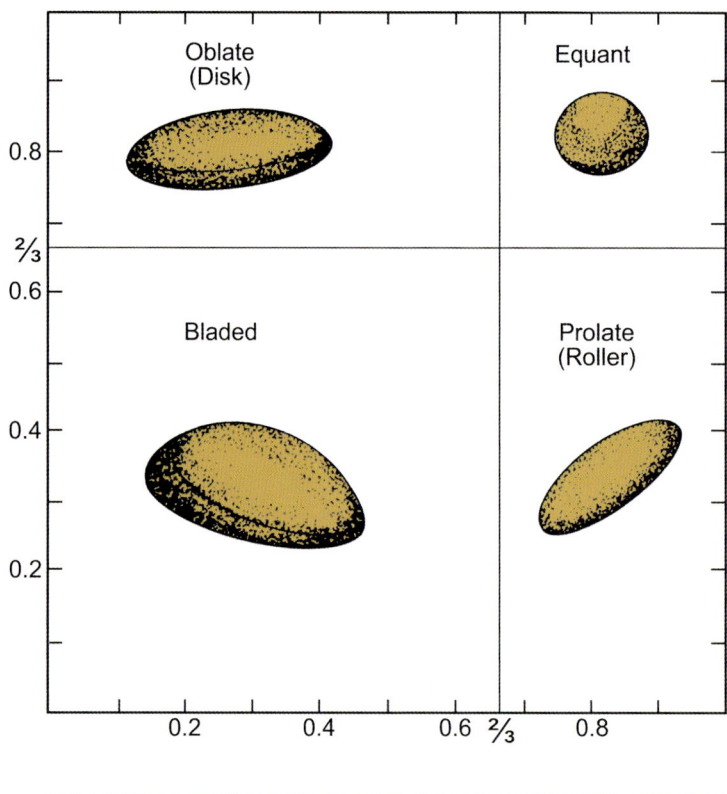

Figure 3.8. Grain-shape classification showing the four common categories. Note that there is a gradual transition between all of them.

Figure 3.9. Close-up of a quartz sand from a Florida Panhandle beach. These grains have a spherical shape typical of that mineral.

understand this is by considering the relative length of three mutually perpendicular axes. When these are the same length, the particle is *spherical*, the most common general shape of beach sediment particles (figure 3.8). Some grains have one long axis, and others have one short axis. Grain shape can reflect the crystallographic nature of the mineral grains. For example, quartz is the most common mineral in beach sand, and quartz has no *cleavage*, planes of crystallographic weakness. As a result it commonly is spherical (figure 3.9). By contrast, some minerals such as mica are platy, but mica is not common in beach sediments.

MATERIALS, STRUCTURES, AND SOURCES 53

Figure 3.10. A beach sediment mixture of whole and broken shells and mineral sand. This is a fairly common situation along the Gulf Coast.

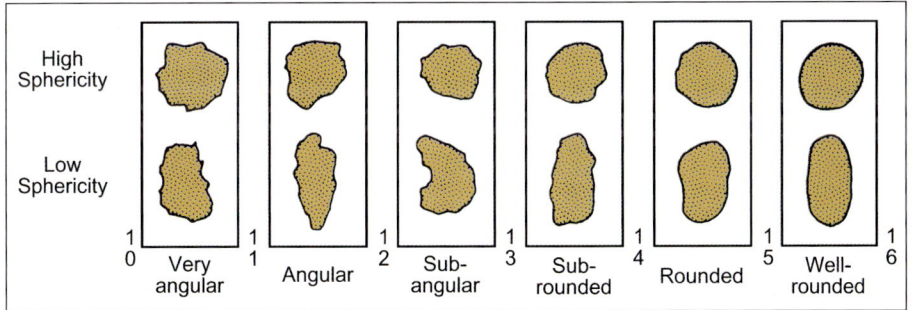

Figure 3.11. Scale of grain roundness. The verbal names are commonly used to describe roundness.

Probably the most varied grain shape of all beach materials occurs in shells or shell debris. The dominant organisms are bivalves and snails, but each has a wide range of shapes (figure 3.10). The broken shells further expand the possibilities of shape. These biogenic particles are transported in waves and currents in very different ways than the typical more spherical quartz and other mineral grains. For this reason we find shells and large shell fragments mixed with sand grains.

The *roundness* of sediment grains is commonly used to interpret beach activity. This is the smoothing of the edges of the grain; it does not refer to its three-dimensional character. For example, a hot dog is well rounded but not spherical, whereas a cube is rather spherical but is very angular, not at all rounded. Roundness can be interpreted to help understand the history of the sediment particles on the beach. Generally, roundness is the result of transportation and/or high energy during which particles are abraded, resulting in increasing roundness. This grain attribute can be expressed verbally or quantitatively (figure 3.11).

We spend time studying the texture of beach sediment, as well as sediment from other coastal environments, because these characteristics allow us to make some interpretations about the beach environment. Additionally, people who study ancient sediments and rocks typically want to be able to interpret the nature of the environment of deposition in which samples that they are studying were deposited. To that end we can make some generalizations about interpreting beach sediments. They are typically well sorted because they are subjected to relatively high-energy conditions of the waves and wave-generated currents. The backbeach is also well sorted and rounded because it has passed through the active beach and is now deposited by the wind. These fluid transport processes tend to sort particles into a narrow range of sizes. If two types of sediment are present, such as shells and sand, then the bimodal sediment does not display good sorting. Sediments of the nearshore/surf zone also display a similar set of textural characteristics.

Mineral Composition

Although most beaches, especially those along the Gulf Coast, are composed mainly of quartz, there is an infinite list of possible minerals that can be in beach sediments. Most of the sediment is *terrigenous*, derived from land. In fact, this sediment is the product of rocks that were eroded from distant locations, were then transported by rivers to the coast, and may have been there for many thousands, perhaps millions, of years. These sediment particles are relatively inert chemically at surface conditions, so they are quite stable. The main components of coastal sediments are quartz, feldspar, and rock fragments. If we extend our scope into estuaries or river deltas, clay minerals are also abundant. The term *clay* may be confusing because there are clay minerals and a grain-size classification that is clay size. The latter is <1/256 mm or about 4 microns, well below the size that we can see, even with a decent light microscope.

In addition to the above-mentioned abundant minerals in coastal sediments, *accessory minerals* are present in the original source rocks as minor

MATERIALS, STRUCTURES, AND SOURCES   53

Figure 3.10. A beach sediment mixture of whole and broken shells and mineral sand. This is a fairly common situation along the Gulf Coast.

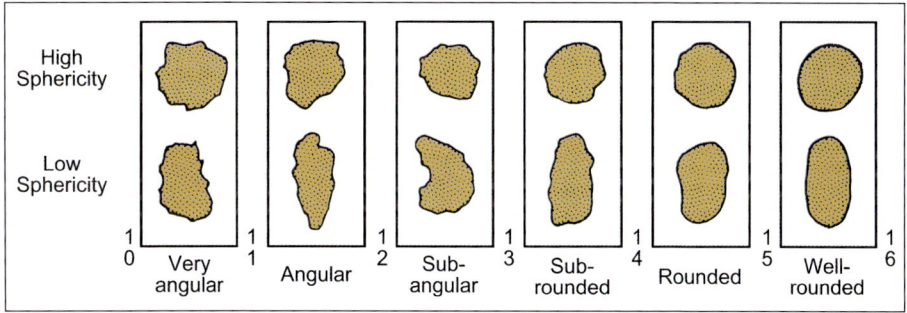

Figure 3.11. Scale of grain roundness. The verbal names are commonly used to describe roundness.

Probably the most varied grain shape of all beach materials occurs in shells or shell debris. The dominant organisms are bivalves and snails, but each has a wide range of shapes (figure 3.10). The broken shells further expand the possibilities of shape. These biogenic particles are transported in waves and currents in very different ways than the typical more spherical quartz and other mineral grains. For this reason we find shells and large shell fragments mixed with sand grains.

The *roundness* of sediment grains is commonly used to interpret beach activity. This is the smoothing of the edges of the grain; it does not refer to its three-dimensional character. For example, a hot dog is well rounded but not spherical, whereas a cube is rather spherical but is very angular, not at all rounded. Roundness can be interpreted to help understand the history of the sediment particles on the beach. Generally, roundness is the result of transportation and/or high energy during which particles are abraded, resulting in increasing roundness. This grain attribute can be expressed verbally or quantitatively (figure 3.11).

We spend time studying the texture of beach sediment, as well as sediment from other coastal environments, because these characteristics allow us to make some interpretations about the beach environment. Additionally, people who study ancient sediments and rocks typically want to be able to interpret the nature of the environment of deposition in which samples that they are studying were deposited. To that end we can make some generalizations about interpreting beach sediments. They are typically well sorted because they are subjected to relatively high-energy conditions of the waves and wave-generated currents. The backbeach is also well sorted and rounded because it has passed through the active beach and is now deposited by the wind. These fluid transport processes tend to sort particles into a narrow range of sizes. If two types of sediment are present, such as shells and sand, then the bimodal sediment does not display good sorting. Sediments of the nearshore/surf zone also display a similar set of textural characteristics.

Mineral Composition

Although most beaches, especially those along the Gulf Coast, are composed mainly of quartz, there is an infinite list of possible minerals that can be in beach sediments. Most of the sediment is *terrigenous*, derived from land. In fact, this sediment is the product of rocks that were eroded from distant locations, were then transported by rivers to the coast, and may have been there for many thousands, perhaps millions, of years. These sediment particles are relatively inert chemically at surface conditions, so they are quite stable. The main components of coastal sediments are quartz, feldspar, and rock fragments. If we extend our scope into estuaries or river deltas, clay minerals are also abundant. The term *clay* may be confusing because there are clay minerals and a grain-size classification that is clay size. The latter is <1/256 mm or about 4 microns, well below the size that we can see, even with a decent light microscope.

In addition to the above-mentioned abundant minerals in coastal sediments, *accessory minerals* are present in the original source rocks as minor constituents and are quite stable. As a result, they are transported along with the major erosional components and accumulate with them on the beach or in other coastal environments. Actually, the beach sediments commonly include a wide range of mineral components (table 3.2). Many of the accessory minerals are commonly called *heavy minerals* because their specific gravity (density) is high, relative to the common coastal mineral grains. Quartz, feldspar, rock fragments, and shell material (calcium carbonate) are commonly about 2.7 gm/cm^3. The so-called heavy minerals are generally >2.85 gm/cm^3. The accessory minerals generally make up less than 1% or 2% of the beach sediments.

These accessory minerals can help us interpret the recent conditions on the beach. Because they have a relatively high specific gravity, they behave differently than the lighter, common beach materials. This is especially the case during relatively high-energy conditions of wave activity. As sediment is removed from the beach by waves and swash, these heavy mineral grains tend to be left behind and the lighter common grains are removed. The result is often a thin layer of dark, heavy minerals (figure 3.12). These thin layers are indications of recent erosion activity on the beach. If a trench is dug across the foreshore, sometimes we can see many of these layers (figure 3.13). A steep slope of these dark layers overlain by more typical gently sloping layers indicates that there was a storm (the steep, dark layers) followed by a period of sediment being added to the beach (gently sloping layers of typical beach sediment) (figure 3.14).

Biogenic Minerals and Sediment Grains

There are many organisms along the coast that contribute to the overall sediment budget. The skeletal material of these organisms comprises calcium carbonate in two different mineral forms: aragonite and calcite, which have the same chemical composition but slightly different crystallography. Oysters are the most common of the skeletal invertebrates that have calcite shells. Most other invertebrate shells are aragonite. This is the case from single-celled foraminifera to the largest snails or coral heads. Some of these organisms are floaters (plankton), and others are swimmers (nekton), but most live on the sediment surface itself (benthos).

The skeletal remains of these organisms can be a

TABLE 3.2. MEAN DENSITY OF SOME MINERALS FOUND IN BEACH SANDS

Mineral	Mean density (gcm^{-3})
Aragonite	2.930
Augite	3.400
Calcite	2.710
Foraminifera shells	1.500
Garnet	3.950
Hornblende	3.200
Magnetite	5.200
Microcline	2.560
Muscovite	2.850
Orthoclase	2.550
Plagioclase	2.690
Quartz	2.650
Rutile	4.400
Zircon	4.600

Figure 3.12. A thin, dark layer on the beach indicating a period of high energy and some erosion. The erosional scarp testifies to the recent erosion.

Figure 3.13. Trench in the beach showing multiple thin layers of dark minerals.

Figure 3.14. Trench across the foreshore beach showing a thick, steep heavy mineral layer that indicates erosion overlain on the left by typical beach sediment that accumulated after a storm.

significant contributor to coastal sediments, in some places 100%! It is common for them to be broken in the surf zone before they reach the beach. Some of the shells are quite durable, and others are not. The thickness and shape of the shell are important factors in their strength. Some of the organisms have skeletons that become disarticulated shortly after the organism expires. The most common shell types in beach sediment are bivalves and snails, with foraminifera a distant third.

It is common for the skeletal material to be physically broken into sand-sized particles, but it is also common for larger shell pieces or even whole shells to be incorporated into beach sediment (figure 3.15). Some beaches in south Florida, Cuba, and southern Mexico are composed entirely of skeletal material.

Some organisms have siliceous (quartz) skeletal material. Sponges have microscopic elongate components called spicules in their soft tissue, and diatoms tend to be made of silica. Although this silicate material is quite hard and chemically inert, the individual particles are very small and the contribution of both sources combined is much less than 1% of the sediment along the coast.

◀ Figure 3.15. Mixture of skeletal material and terrigenous particles that compose a beach sand.

▲ Figure 3.16. Ooids, which are spherical, medium sand-sized grains of calcium carbonate, are common in low latitudes where carbonate sediments dominate, but they are rare in the Gulf.

Chemical Sediments

Although uncommon, some chemically precipitated particles can locally be important as beach sediment: calcium carbonate grains that are a nucleus with thin layers precipitated around it to form *ooids*. These are medium, sand-sized, spherical grains that are very common on beaches and sand shoals in the Bahamas (figure 3.16). They also occur in the Gulf of Mexico in the Yucatán area and locally in the Florida Keys, Marquesas Keys, and the Dry Tortugas. Ooids are limited to low-latitude areas where carbonate sediments dominate.

The hardness of mineral grains is an important factor in the durability of erosion products and the behavior of sediment particles in the high-energy surf zone and beach environment. Minerals are ranked 1 to 10 by hardness on the Moh's Hardness Scale. Diamond is the hardest at 10, and talc is the softest at 1. The common constituents of beach sediments fall somewhere in between. In fact, two hardness values dominate beach sediments. Most of the particles, the quartz, feldspar, and rock fragments, are in the 6–7 range of hardness. Calcium carbonate is only 3. Additionally, calcium carbonate is easily weathered chemically, whereas the others are nearly inert. These data mean that the skeletal material has a finite "life span" on the beach, whereas the terrigenous particles can last essentially forever.

Figure 3.17. River with sand-point bars, an environment where erosion particles may spend some time on the way to the coast.

Sources of Sediment

All of the nonskeletal minerals just discussed in detail are derived from preexisting rocks. The potential compositional spectrum is very wide. In order for these sediment grains to be removed and transported to the coast, they must be physically durable and chemically rather inert. Most beach sediment grains comply with these characteristics. These sediment particles may travel more than 1000 km before they reach the coast. They may make many stops along the way in various fluvial environments, particularly point bars (sand accumulations) on a river bend—stops that may last for months, years, decades, or longer (figure 3.17).

Eventually the sand particles reach the coast. Some are incorporated into deltas, and others are discharged into estuaries or directly into the Gulf of Mexico. During the past million or so years the earth has experienced multiple cycles of sea-level change ranging from about 130 m below the present position to about 40 m above it. During these cycles rivers carry sand to various parts of the coastal plain and the exposed continental shelf. The rise and fall of sea level permit waves to rework these sand deposits and develop beaches. The result is that these terrigenous sediments have moved from their origins in rock formations somewhere on the continent to a beach on the Gulf of Mexico. Some of these sediment grains may be hundreds of millions of years old.

Sedimentary Structures

Processes of the surf zone and beach interact with sediments to produce a variety of recognizable features. Most are simply curiosities that are part of the beach environment, but some can be preserved in the ancient rock record and help us understand the conditions of the environment when the sediments were accumulated. These are collectively called *sedimentary structures* and may be produced by water motion, wind, or organisms.

The most widespread and temporally continuous surface structures in the surf zone and beach are *ripples* (figure 3.18). These surface features are regular and equally spaced and are an element of a spectrum of similar features called *bedforms*. The ripples are produced by the circular or oscillatory motion of water particles and have shapes similar to those of miniature waves. These surface features are nearly symmetrical. When combined flow conditions are present, the ripples are distinctly asymmetrical. In the event that the current portion of

Figure 3.18. Underwater photo of typical wave-formed ripples that characterize the sand-dominated nearshore zone.

Figure 3.19. Schematic diagram that shows how ripples vary with changing energy conditions.

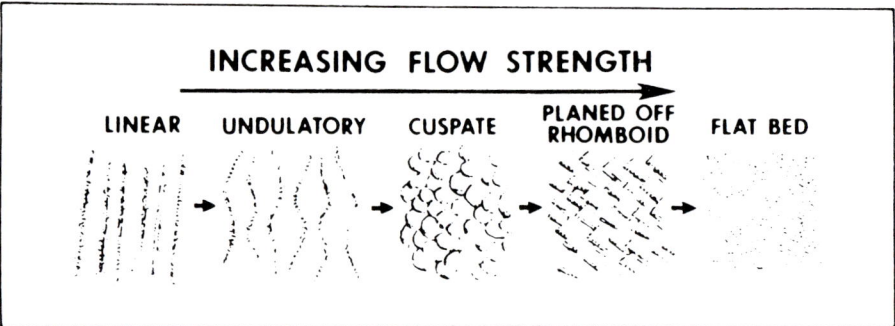

Figure 3.20. Antidunes on the foreshore of a gently sloping beach on the coast of Georgia.

the combined flow dominates the environment, the ripples are smoothed and the sediment bed surface is flat (figure 3.19).

The surface of the nearshore zone where the bar-and-trough structure is present is essentially a blanket of ripples. It is common that the crest of the longshore bars where wave-generated currents are strongest will display a flat

Figure 3.21. Wind-generated ripples on the backbeach surface.

Figure 3.22. Swash marks made by the most landward uprush of a wave.

surface due to combined flow currents and intense turbulence from breaking waves that persist there. The rippled surfaces are generally in continuous flux as waves change size and water depth varies with tidal cycles.

Another type of bedform common on the gently sloping sand foreshore is *antidunes*. These are current-generated surface features produced during the backwash of waves as gravity returns water after wave runup. They are gently undulating features on this part of the beach environment (figure 3.20). They form best when the sediment is fine, well-sorted sand. It is also common for ripples to form on the backbeach where dry sand conditions prevail. These wind-generated ripples display a distinctly asymmetrical profile (figure 3.21). They are commonly low relative to wave-generated ripples. The backbeach ripples are oriented with the wind direction.

Figure 3.23. Current crescents formed by local water currents around an obstacle.

Figure 3.24. Sand shadows on a dry beach where sand accumulates on the downwind side of an obstruction.

Swash marks are very delicate, curved accumulations of sand forming arcuate lines in the active swash zone (figure 3.22). These are formed at the position of maximum wave runup where water flow reverses, leaving a very small amount of sediment. The shape of the swash marks is an indicator of the wave energy. The more arc-shaped they are, the lower the wave energy. Those that are nearly straight suggest relatively high wave energy. The shoreline morphology, such as cusps, can also influence the pattern of swash marks.

Other common surface structures are the result of currents moving over pebbles or shells and producing direction-oriented structures. There are two kinds of such structures: one produced by water currents and the other produced by wind. *Current crescents* are parabolic-shaped excavations made by water currents around an obstruction (figure 3.23). The open end of the parabola is downcurrent. These features are ephemeral but have been found in the ancient rock record. They are even more common in streams than on the beach. The wind-generated features, *sand shadows*, occur where a small amount of sand accumulates on the downwind side of an obstruction (figure 3.24). These features are also quite ephemeral and common in the desert environment.

SUGGESTED READING

Brogdon, D. R. 1954. "Beach Sands of the Gulf Coast—Tamaulipas, Mexico." Master's thesis, University of Texas, Austin.
Brooks, G. R., L. J. Doyle, R. A. Davis, N. T. DeWitt, and B. G. Suthard. 2003. "Patterns and Contents of Surface Sediment Distribution: West-Central Florida Inner Shelf." *Marine Geology* 200:307–24.
Hayes, M.O. 1965. "Sedimentation in a Semiarid, Wave-Dominated Coast (South Texas) with Emphasis on Hurricane Effects." PhD diss., University of Texas, Austin.

4

Human Impact on Gulf Beaches

MOST of the world population lives within an hour's drive of a beach. The influence of humans on the coast has been extensive and intensive, and it will continue in the future. The entire coastal system has been impacted by various human activities: the dunes, estuaries, tidal inlets, and most certainly, the beaches. This discussion includes the spectrum of human influence on the beaches going back to some of the early efforts to protect and/or control coastal change. Since the 1960s we have made changes in how we manage the coast, including the beaches. These changes have been aimed at being less intrusive into coastal dynamics and have provided more aesthetic methods for beach management.

Human efforts to control some of the changes that beaches experience focus on coastal erosion and inlet management. There have been numerous approaches to these efforts, some that work pretty well and others that definitely do not. The US Army Corps of Engineers has led the way in the effort to eliminate or moderate beach erosion problems. They have taken considerable criticism over the years because of their approaches to coastal management. Most recently the Corps, as it is commonly known, has moderated its approach and the public has been appreciative of their efforts. Now all Gulf Coast states also have agencies that are responsible for coastal management and for regulating various activities there, especially construction. Typically it is necessary to obtain permits for any type of coastal modification from both the federal (Corps) and state government agencies. The current system is not perfect, but it works much better than in the past, and the coast has benefited greatly from this cooperation.

Shore-Parallel Structures

Beach erosion is widespread and has been going on ever since beaches have existed. It has been a "problem" only since human development has led to construction along the coast. This activity has placed a price tag on beach erosion as houses and commercial buildings have been damaged or destroyed. The first reaction to such events is to protect the property from further erosion. A wide range of protective measures has been used, some helpful and others not.

Placing something along the problem shoreline has been the first line of defense. Such efforts have been quite varied over time with the cost of protection being a significant limiting factor. The most primitive approach is to place some type of heavy material at the point of erosion and hope for the best. Large boulders of various rock types have been used (figure 4.1). In some locations they have simply been piled along the erosion scarp without any attempt at organization. In other examples, each piece of rock is carefully placed to provide a very stable rock wall (figure 4.2). This type of material is commonly called *riprap* and represents one of the least expensive approaches to shoreline protection.

Some locations have experienced rather absurd methods of protection. The eastern coast of Lake Michigan has experienced erosion of high bluffs upon which sit residential developments. In the late 1960s and early 1970s, junk cars were placed on these bluffs to prevent further erosion (figure 4.3). This technique was illegal, but it took place before strict rules of shoreline protection were adopted.

Another approach is to use geotextile tubes, very large plastic tubes placed strategically along the shoreline and pumped full of sand to produce a substantial barrier to prevent further erosion (figure 4.4). They are designed and placed to eventually be covered with sand, thus presenting an aesthetic coastal condition. Problems do occur with vandalism (cutting the bags) and storms that cause the bags to break. However, when this happens, there are methods of patching small tears or cuts, and if broken badly, the clothlike material can be extracted and discarded, thus returning the beach to a natural condition.

In the past, before permits were required, people put whatever was available at the point of erosion at some locations, including tree limbs, broken-up concrete, or other aesthetically unpleasing materials. Cost was a major factor, and most of this type of protection was done by an individual property owner or a small group of owners.

Figure 4.1. Boulders along the backbeach to help stop erosion.

Figure 4.2. Protection provided by well-organized blocks of granite along the Texas coast at Sargent Beach.

68   GENERAL CHARACTERISTICS AND DYNAMICS OF BEACHES

Figure 4.3. Junk cars used to protect bluffs from eroding along the coast of Lake Michigan.

Figure 4.4. Geotextile tubes filled with sand provide some protection from beach erosion.

Seawalls

Probably the most common method for protection from beach erosion is construction of seawalls. These structures range widely in materials and configuration. A big factor in construction is cost. Seawalls may be built in front of individual properties or may extend for long distances in front of entire communities. The rules and regulations for these structures have changed greatly over the years. As is commonly the case, permits have been required for the past few decades and the requirements for successful application for such permits have become strict, making them very difficult to obtain. The following discussion considers a wide spectrum of these structures, including some that are not present on the Gulf Coast.

The problem with seawalls is that they do not absorb any wave energy. They reflect it, much like a mirror reflects light. As a consequence, eventually there will be fatigue and the materials will break down. Another problem concerns scour at the base of the wall because of the reflection of wave energy. Some walls are not constructed deep enough below the beach surface, and scour will undermine the wall and cause it to fail.

The most common type of seawall is a vertical structure that may be constructed of various materials. The least expensive and also the least durable are wooden planks placed vertically (figure 4.5). These may be back-filled to provide some additional resistance to wave attack. The type of material used is a major limiting factor in the performance of vertical seawalls.

A much more durable type of vertical wall is built from sheet piling, which may be metal, vinyl, or fiberglass and is much more costly than wood (figure 4.6). These materials can be pounded or jetted deep into the substrate so that scour from waves is not a problem. Such structures are typically capped with wood, concrete, or metal to make them more wave resistant and more visually tolerable. Concrete vertical walls are also pretty common, especially where quite long reaches of protection are required (figure 4.7)

Some seawall designs involve a bit of creativity. The objective is obviously to provide protection from erosion, but something must be done to absorb some of the wave energy to minimize the fatiguing of the vertical wall. One technique is to slope the wall by using a steplike design (figure 4.8). This

Figure 4.5. Construction of a seawall of wooden planks, an inexpensive but short-lived version.

▲▲ Figure 4.7. Poured concrete seawall on Sand Key, Florida.

▲ Figure 4.9. Close-up of seawall built in the early 1900s at Galveston, Texas.

▲▲ Figure 4.6. Seawall constructed using sheet piling that can be made of metal, vinyl, or fiberglass.

▲ Figure 4.8. Seawall with a stair-step design to help absorb wave energy.

approach is expensive, but it does accomplish the necessary objectives of durability and protection along with dissipation of wave energy. Probably the most famous seawall on the Gulf Coast, that at Galveston, Texas, has a curved surface to permit waves to "run up" the face of the wall somewhat and thereby absorb and spread the wave energy over time and space (figure 4.9). This structure was built shortly after the devastating hurricane of 1900. It has lasted more than a century and is still in good shape. The engineering community commonly calls it one of the seven wonders of the engineering profession.

Another practical approach to erosion protection is to combine riprap with a vertical seawall (figure 4.10). This is expensive, but the old adage of "you get

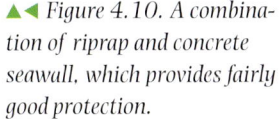

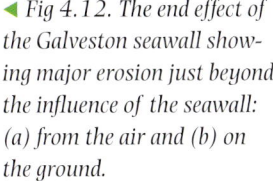

▲◀ *Figure 4.10. A combination of riprap and concrete seawall, which provides fairly good protection.*

▲ *Figure 4.11. Tie-downs, which are important stabilizing factors, shown behind the seawall.*

◀ *Fig 4.12. The end effect of the Galveston seawall showing major erosion just beyond the influence of the seawall: (a) from the air and (b) on the ground.*

what you pay for" is certainly applicable here. The riprap absorbs most of the wave energy that impacts the structure, and the seawall behind it is protected and also protects the property landward of it. Most seawalls are now fronted by beach sand, natural in some places and placed as nourishment. In this format the sandy beach protects the coast, and the seawall serves as the last line of defense of the upland property.

One method of construction commonly required for seawalls is the use of *tie-downs*. These are steel rods embedded into the wall and anchored landward at an angle with a "dead man" weight (figure 4.11). This type of construction prevents the seawall from falling seaward in the event that it is undermined by scour at the base. A major problem associated with seawalls is the "end effect." These structures have a finite extent along the shoreline. At the end there is usually significant erosion where the structure is absent (figure 4.12). In the case of local structures this can cause severe problems for properties adjacent to the wall, either on the end or between two local walls. Major shoreline retreat can take place.

Breakwaters

There are also many shore-parallel structures that are seaward of the dry beach, at least when they are constructed. These structures are designed to reduce wave action at the shoreline and thereby reduce or prevent erosion of the beach. The design and placement of these structures are extremely important, and they require considerable research prior to formulating a plan. The wave climate must be determined by monitoring wave size and direction. It is also necessary that sediment transport, both shore-parallel and shore-normal, be determined and studied before making a decision on the location and extent of these offshore structures.

In some situations a *salient* will develop between the structure and the original shoreline. This is an accumulation of sediment that results from lower wave energy and the inability to keep longshore sediment transport moving along the beach (figure 4.13). In most cases, the breakwaters are placed at a substantial distance from the beach to avoid the problems presented by salients. Some situations call for several of these structures, such as on the coast of Louisiana where the small barrier islands are in jeopardy (figure 4.14).

Not all breakwaters are just parallel to the shoreline. Some have multiple parts oriented in other directions. These are typically associated with some type of harbor and are designed to provide safe mooring for boats. The idea is to keep wave energy from passing into the anchorage. It is not common for such structures to be adjacent to an open beach along the Gulf of Mexico coast.

HUMAN IMPACT ON GULF BEACHES 73

◀ *Figure 4.13. Small breakwater on Sand Key, Florida, where the salient has completely filled in between the structure and the beach. It took less than 10 years for this to take place.*

Figure 4.14.
◀*(a) A very long breakwater and*
◀▼ *(b) multiple breakwaters along the Louisiana coast. Photos courtesy G. Stone.*

(a)

(b)

Shore-Normal Structures

There is also a range of hard structures that are constructed perpendicular to the shoreline and designed to stabilize the beach by trapping sediment and controlling the impact of wave energy on the beach. Such structures may occur in groups or as individuals. Longshore transport of sediment is a natural process sometimes called "the river of sand." The rate of sediment transport may be very high, and in most cases it is not a desirable condition for maintaining a stable beach. These shore-normal structures are intended to increase beach stability by interrupting the river of sand carried by longshore currents.

Groins

The longshore transport of sediment along the beach can be a serious erosional process. This is especially true on coastal beaches that experience a relatively large volume of net sediment movement in one direction or the other along the beach. Many coasts experience this phenomenon, and for many decades coastal management entities along the Gulf Coast have tried to mitigate this situation. The approach is to construct some type of structure from the beach into the surf zone that will capture sediment and keep the longshore transport to a minimum. The annual net rate on the Gulf Coast tends to be less than 100,000 m^3 with most locations being less than 50,000 m^3. On the Atlantic coast of Florida near Jacksonville, the rate is about 500,000 m^3 per year.

Groins may be constructed of geotextile tubes, wood pilings, metal pilings, concrete, or so-called concrete dog bones (figure 4.15). The nature of the material used depends on permit requirements and cost. Typically there are many short groins along a beach (figure 4.16). The length and height of each of the structures are very important to the success of the groins. Unfortunately, most do not perform very well.

If properly designed and located, each structure should trap sediment up to the elevation of the crest and additional sediment in transport will pass over it (figure 4.17). Eventually all of the groins should be buried. If this happens, the structures have been quite successful (figure 4.18). More commonly, there is a downdrift effect on each of the groins. Because overdesign is a common problem with these groins, they trap too much sediment, and erosion occurs in the lee of the structure (figure 4.19).

Groins on the Gulf Coast tend to be rather small, but in other locations they

▲◀ *Figure 4.15. A series of dog-bone groins on Captiva Island, Florida. The structures are cast concrete and are placed in an irregular line out from the beach.*

▲ *Figure 4.16. Aerial view along Sand Key in Pinellas County, Florida, showing a typical groin field.*

are very large. This is particularly the case on the North Sea coast of the Netherlands and Germany. Here the wave energy and the rate of longshore sediment transport are high. These structures are very carefully built from large masonry blocks. These countries have been constructing such structures for centuries, and their design and placement rarely result in significant downdrift erosion. Generally, distribution of sediment is rather uniform on both sides (figure 4.20).

Another type of groin, a *terminal groin*, is located at the end of a coastal reach, commonly a barrier island (figure 4.21). These structures are designed to hold sand on the beach and not permit it to be transported into an adjacent channel. The objective is twofold: keep more sand on the beach and prevent sand from causing navigation problems in an adjacent inlet channel. They are constructed of hard materials: concrete, steel sheet piling, or well-placed rock material.

Jetties

Jetties are also perpendicular to the shoreline, but they stabilize tidal inlets, and much like terminal groins, prevent sediment from moving along the shore. The typical jetty is constructed of the same material as groins, although geotextile tubes are not often used. These structures are generally much longer than any groin type and extend into navigable water. The objective of building jetties is to prevent closure of tidal inlets or clogging of the inlet channel with sediment that would render it unnavigable. In many locations the jetty design

Figure 4.17. A small groin that has been designed correctly and is performing properly. At this stage sand will pass over the crest of the structure, and there should not be any downdrift erosion.

Figure 4.18. A groin field that has been quite successful on the central Florida coast. It is surrounded and nearly buried by sand.

fails to do the job. There are places where the rate of longshore sediment transport is so large that the fillet of sand next to the jetty is full and sand passes by the end of the structure into the channel (figure 4.22). Another problem is that many of these structures are constructed of only boulders, which leak sediment and do not solve the problem of sand clogging the channel.

One of the very common problems with jetty structures at inlets is that

Figure 4.19. Downdrift erosion of a groin that is too large and traps too much sediment (dark line indicates the groin).

Figure 4.20. Large groin structure on the North Sea coast showing rather uniform accumulation of sediment on both sides. This situation is the result of excellent design of the structure and its compatibility with the coastal processes.

Figure 4.21. Terminal groin at the northern end of Treasure Island on the Florida Gulf Coast. The structure is trapping sediment, but it is too short and sand is moving into the inlet, causing navigation problems.

Figure 4.22. Sediment accumulation at Port Mansfield jetties in South Texas showing difference in sediment transport rates from each side. Photo courtesy US Department of Agriculture.

they interrupt the transport of sand along the coast. As a consequence, erosion occurs adjacent to the downdrift jetty on the opposite side of the channel (figure 4.23). This erosion may be severe and result in damage or loss of infrastructure or buildings. Even though jetties are commonly a problem, they are necessary for navigation. Some do show good performance with a balance of sediment on both sides of the inlet without interfering with flow through the

Figure 4.23. Extensive erosion downdrift of a jettied inlet, a very common situation. This example is at Upham Beach on the central Florida coast.

HUMAN IMPACT ON GULF BEACHES 79

Figure 4.24. Jetties that are doing a good job of keeping the channel free of sediment: (a) with no offset of the shoreline and (b) with an offset.

Figure 4.25. Sediment by-passing facility to eliminate the problem. There are few of these installations around the coast, very few of which function well. On the Gulf Coast there has been one small test installation (a) at Mexico Beach on the Florida Panhandle. It was active in the mid-1970s. The only sizable operating system of this type is (b) at Indian River, Delaware. Courtesy Delaware Department of Natural Resources.

channel (figure 4.24a), but some will dam longshore sediment transport and still not cause navigational problems (figure 4.24b).

To accommodate the problems of inlet structures trapping large volumes of sediment, coastal engineers have tried two general approaches. One is to intentionally allow sediment to pass through a constructed low portion of the jetty on the updrift side of the channel. This permits sediment to move into the channel where currents are relatively strong and may keep the sediment from accumulating. If sediment is allowed to pass to the end of the jetty and move around it to the channel, the currents are commonly not strong enough to keep the sediment from accumulating in the channel. This is not a foolproof approach, and dredging of the channel is commonly necessary.

Another approach is to install a transfer station that will move sediment from the fillet on the updrift side of the inlet to the downdrift side. It sounds like a good approach, but few of these have been installed and even fewer have been successful. Sediment is sucked into a large transfer pipe and pumped across the inlet and onto the beach downdrift of the other jetty (figure 4.25a). These installations are expensive, and the problems of plumbing and pumping in a seawater environment are extensive and severe. The only sizable example of a successful pump transfer system is at Indian River in Delaware (figure 4.25b).

Beach Nourishment

Beginning in the late 1960s and early 1970s, hard structures were being replaced by soft methods of shoreline protection. The primary approach was *beach nourishment*, a technique that places sediment on the existing beach, essentially to build a new one. The beach is the primary tourist attraction on the coast, especially on the Gulf Coast where weather is generally good year-round. Beach nourishment is a complicated and expensive approach for shoreline protection and is a temporary solution to a long-term problem. Some locations perform very well, and others not so well (figure 4.26); nevertheless, it is popular and very cost effective.

The basic approach is to locate a source of sediment as similar as possible to the sediment that naturally exists or existed on the beach to be nourished. The sediment is then dredged and placed on the eroding beach according to specifications stated in the permits that have been granted. Funding is typically cost-shared between the federal, state, and local governments, commonly in a decreasing percentage. Permits are required by the Corps of Engineers and any appropriate state agencies as well as agreement of local jurisdictions. The use of public funds to nourish a beach requires that public access is provided, generally with location spacing required by a formula based on the source(s) and amount of funding.

Beach nourishment has been most common in both space and time on the Gulf Coast of Florida, partly because Florida is such a tourist-dominated economy and has a large coastal population. Beach erosion is common along much of this coast. Another reason is that the Gulf Coast of the Florida peninsula is a sediment-starved coast because almost no sediment is carried there by rivers and most of the inner shelf is floored in limestone.

Once it is determined that a nourishment project is needed, then it is necessary to locate the best and closest source of a sufficient amount of appropriate sediment. There are two sediment environments along the coast where this material is typically located. The best is commonly the ebb-tidal deltas at the seaward mouth of tidal inlet channels. These sediment bodies have nourishment material that comes close to matching that of the beach, with the only typical deviation being a larger amount of shell debris. They are especially common on the Florida coast. Some difficulty can occur in obtaining permits for dredging this sediment because removal of large volumes of material can alter the wave patterns in the ebb-delta area and cause problems for performance of the inlet (figure 4.27).

Another common source of nourishment sediment is linear sediment bodies that occur on the inner shelf. These may represent old barrier islands, and therefore their sediment is quite similar to that of modern beaches. These sediment bodies are small off the Florida peninsula and may be far from shore, therefore increasing cost (figure 4.28). They are more common on the northern Gulf Coast from the Florida Panhandle across the Mississippi delta and along the Texas coast. In all situations, the distance from the subject beach is a major factor in the cost of the project.

Once located, the nourishment, or borrow, material must be extracted and moved to the beach being built. This generally takes place in one of three ways: (1) a suction dredge with sediment being pumped or barged to the construction site (figure 4.29a), (2) a hopper dredge that removes the sediment and transports it to the beach (figure 4.29b), or (3) a bucket dredge that removes the sediment and places it in a barge that is towed close to the beach, and then the sediment is pumped onto the beach (figure 4.29c). Hopper dredges draw relatively little water, so they can suck up a load of sand and then move to shallow water, where the sand is pumped directly onto the beach (figure 4.30).

After the borrow material has been dumped on the beach, it is sculpted by large earthmoving machines to a specific profile (figure. 4.31). The design

HUMAN IMPACT ON GULF BEACHES 83

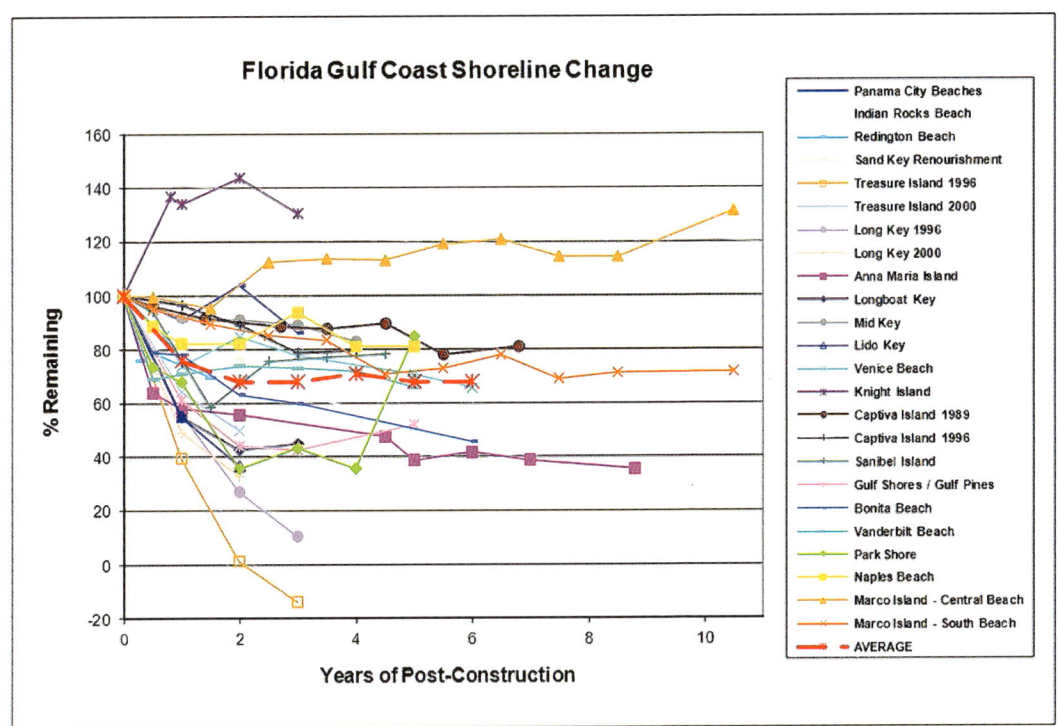

▲ Figure 4.26. *Performance of the many nourishment projects on the peninsular Gulf Coast of Florida. Note that some perform very poorly and some very well. Courtesy Shore and Beach Resources Center, Florida State University.*

◀ Figure 4.27. *Excavation scar for a nourishment project near an inlet along the central Florida coast shown seaward of the heavy line.*

▼ Figure 4.28. *Seismic section showing small sand bodies on the inner shelf off the Florida coast. These might be relict barrier islands and have some potential for beach nourishment. From S. D. Locker, A. C. Hine, and G. R. Brooks, "Regional Stratigraphic Framework Linking Continental Shelf and Coastal Deposits of West-Central Florida," Marine Geology 200 (2003): 351–78.*

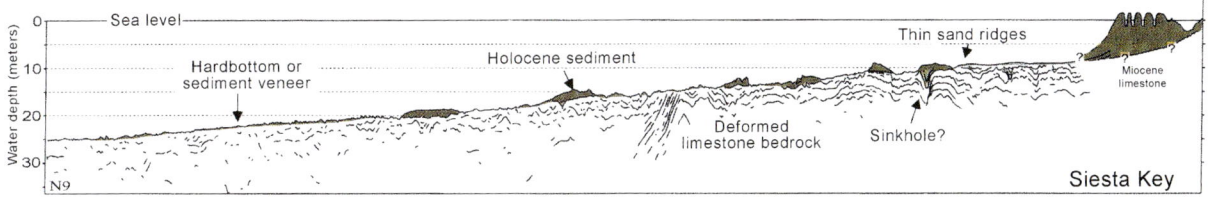

Figure 4.29. Methods for removal of borrow material to be used for beach nourishment include (a) suction dredge, (b) hopper dredge, and (c) drag line and barge. Photo b, courtesy Damen Dredging.

(a)

(b)

(c)

Figure 4.30. Hopper dredge pumping sediment onto beach. Photo courtesy Westminster Dredging.

varies from location to location, but the beach surface is commonly about 1.8 m above mean sea level, and the width of the beach commonly ranges from 33 to 66 m. The foreshore portion of the beach is constructed much steeper than under natural conditions because of the inability of the earthmovers to work in saltwater. The result is a wide and generally beautiful beach (figure 4.32). The surf zone processes rework the foreshore to a much less steep slope, making it appear that the beach is losing sediment due to erosion (figure 4.33), but the beach sand is simply being carried into the subtidal area just seaward of the shoreline. It is even common for the nourished beach to develop a significant erosional scarp during the initial adjustment period.

Cost of these nourishment projects is high. Most require at least a few hundred thousand cubic meters of borrow material, and the cost can be in excess of $25/m^3$. They are multimillion-dollar projects. A major factor in the cost per unit is the transportation of borrow material from the dredge site to the construction site. The distance can be more than 30 km. The best scenario is to be able to pump the nourishment sediment from the dredge directly to the beach. This can be done if the distance is a few kilometers. If it is more, sediment must be dredged into a large barge, towed to a location just offshore of the construction site, and then pumped onto the beach, or a hopper dredge can be used.

Figure 4.31. Large earthmoving machines distributing sand on the beach being nourished.

▲▶ *Figure 4.32. (a) Aerial view and (b) land view of a completed nourished beach on Sand Key, Florida.*

▶ *Figure 4.33. Steep erosional scarp on nourished beach shortly after completion of construction. This is an indication of the beach profile adjusting to natural processes.*

As mentioned, these beach nourishment projects are temporary. The same processes that caused the erosion that necessitated the nourishment continue, so the new beach also erodes. Depending on the level of wave energy, occurrence of severe storms, and the nourishment material, the time before renourishment can be as short as two to three years or up to eight to nine years. Because of the economic impact of tourism, it is easy to justify renourishment on a cost-benefit basis.

Perched Beach

Most beach nourishment projects are rather lengthy, involving several hundred meters to a few kilometers. These are generally municipal projects with

◀▼ *Figure 4.34. (a) Aerial view and (b) land view of a perched beach on the central Florida coast.*

the local contribution coming via a hotel tax. Occasionally private funds are used to nourish a beach, generally only a small project. One such project was constructed by a homeowner's association in front of high-rise condominiums on the Florida coast near Clearwater, where the beach was completely eroded back to the concrete seawall.

The nourished beach is held in place by geotextile tubes placed like groins. Sand was pumped from nearby offshore. The resulting beach was perfect for the intended use by the residents (figure 4.34). The limiting factor is the durability of the geotextile tubes, which are quite vulnerable to storms. Such local projects are difficult to permit because of the potential end effects for the adjacent properties. This particular example is now gone due to the large-scale nourishment of the entire area along Sand Key, the longest barrier on the Florida Gulf Coast.

Figure 4.35. Aerial view of the coast at South Padre Island, Texas. Photo courtesy Tripadvisor.com.

Surf Zone Nourishment

One of the least expensive approaches to beach nourishment is to place the fill sediment offshore of the target beach. At South Padre Island, Texas, there is considerable development, primarily for tourism (figure 4.35). Various types of structures have been used in some locations, but the most widespread method of protection has been beach nourishment. One of the early projects was completed using material from the Brownsville, Texas, ship channel, which needs maintenance dredging on a regular basis.

The dredge pumped the borrow material from the channel, and it was delivered to the site and placed several hundred meters offshore in the subtidal environment. The objective was to have waves and wave-generated currents transport the sediment onto the beach. Once there, it would behave like a beach recovering from storm erosion. This approach worked quite well, and after several months the beach received most of the nourishment and took on the profile of an accreting beach. The cost of this project was considerably less than the more common method of pumping sediment onto the beach and then sculpting it to match the desired profile.

Dune Enhancement

Although dunes are not strictly part of the beach, they are intimately related to the beach environment. The beach provides the sediment that forms dunes as onshore wind blows over the dry beach. Because the dunes are major protection for the upland area on their landward side due to their high elevation, it is critical that coastal management programs include aiding the dunes in their growth and development.

A basic approach to this effort of dune enhancement is placing the material raked or scraped from the beach surface on the front of the foredunes. This may be flotsam such as the algae *Sargassum* or other species plus the sand itself. Most managers do not consider this to be good practice because it removes sediment and materials that themselves help stabilize the beach. It is typically done to provide a cleaner beach for tourists.

Figure 4.36. Snow fences placed along the backbeach to trap sand and help build and maintain dunes: (a) long continuous fences, and (b) short, en echelon *sections of fence.*

"Snow" fences designed to trap sediment by causing disturbance to the wind motion are quite common. The approach is to place them either in an extended line at the foot of the foredune or in short *en echelon* rows in the same location (figure 4.36). The objective is to bury the fence in sand. On South Padre Island in the undeveloped area, dunes are being stabilized and helped to grow by using large, cylindrical hay bales, and near Surfside, Texas, Christmas trees are used (figure 4.37).

Figure 4.37. Aid in growth and development of dunes can be provided by (a) placing Christmas trees in rows along the backbeach and (b) scattering large bales of hay in the backbeach to attract and stabilize sand in the form of dunes (arrows indicate hay bales).

SUGGESTED READING

Douglas, S. L. 2002. *Saving America's Beaches: The Causes of and Solutions to Beach Erosion.* Hackensack, NJ: World Scientific.

Nordstrom, K. F. 2008. *Beach and Dune Restoration.* Cambridge: Cambridge University Press.

5

Common Animals and Plants of the Gulf Beaches and Surf Zone

ALTHOUGH the beach and surf zone are very dynamic, they do have a community of organisms that is pretty similar throughout the Gulf Coast. Both the plants and animals must be adapted to fairly rigorous conditions: tidal fluctuations, wave attack, wind, little available freshwater, and predators. These conditions limit the diversity of organisms. This discussion does not consider the extremely mobile animals such as birds or fish. The emphasis is on the few common benthic organisms of the beach and surf zone, both mobile and sessile (permanently attached). Others not mentioned here are described in the many books on beach fauna and flora.

Nearshore / Surf Zone

The shallow nearshore zone where waves break and currents can be strong presents a difficult set of conditions for bottom-dwelling organisms. The occasional bivalve or snail may find a place to burrow here to be protected from the typical waves. Epifaunal organisms, which live on the sand surface, are not common due to the wave energy and the mobile substrate. It is important for the wader to be careful of burrowing snails such as *Turritella* and *Oliva*, both of which can put a hole in your heel if you step on them. The other creature that can cause injury is the sting ray (figure 5.1). This animal has a stinging barb that can penetrate the foot or heel. The so-called sting ray shuffle is the way to avoid the problem. When walking through the surf zone, it is best to shuffle your feet, thus warning the ray of your approach and sending it on its way.

One of the most common surf zone animals is the sand dollar (*Mellita quinquiesperforata*). These small echinoderms are about 6–8 cm in diameter (figure

Figure 5.1. Sting rays are very common in the nearshore and sometimes appear in schools of many individuals. Photo courtesy Alex Edgil, University of Alabama, Birmingham.

5.2). They bury themselves partially in the sand to maintain their stability. These creatures do not like to be in intense wave activity. Storms commonly wash many of them onto the beach.

Foreshore

Some animals are typical of the foreshore beach, especially in the swash zone. The surf clam (*Donax variabilis*; figure 5.3), also known as coquina, may

*Figure 5.2. Small sand dollar (*Mellita quinquiesperforata*), which is very common in the nearshore zone where there may be dozens per square meter: (a) the bottom from which it feeds and (b) the top. Photo a, courtesy Gulf Specimen Marine Lab; photo b, courtesy J. W. Tunnell.*

Figure 5.3. The swash zone commonly contains (a) many surf clams (Donax variabilis) that are typically present in (b) a wide range of colors and patterns almost as varied as fingerprints. Photo a, courtesy J. W. Tunnell; photo b, courtesy Fabio Moretzsohn.

be present in hundreds of individuals per square meter. These small bivalves come in a wide range of colors and patterns. They can be exhumed by swash action but will burrow into the sand in a matter of seconds. Their small siphons filter particulate organic debris from the water as their sustenance. These filter feeders have some value to humans as a tasty morsel. In some places, especially southern Spain, the coquina clams are steamed and served with garlic butter as an excellent appetizer.

The other common organism in the intertidal foreshore is the ghost shrimp (*Callianassa major*; figure 5.4a). This animal burrows quite deep into the intertidal beach and constructs a substantial burrow. It feeds by pulling water into the burrow and extracting its nourishment from suspended particulate organic matter. These shrimp produce diagnostic fecal pellets that commonly surround the small opening of the burrow (figure 5.4b). Ghost shrimp are not common along the Florida coast but are abundant on the Texas coast largely because of the relatively plentiful fine suspended material there. The water is less turbid in Florida and east of the Mississippi delta in general. These shrimp can be extracted from the sediment by a homemade suction pump of PVC tubing. They are excellent bait for surf fishing on the Texas coast.

Just at the landward edge of the foreshore can be found the Portuguese man-of-war (*Physalia*). These jellyfish float in the Gulf and commonly wash ashore and onto the beach. They are easy to recognize because of their pink-purple color (figure 5.5). These jellyfish are quite dangerous, and their tentacles can sting with a very powerful poison that will require the victim to seek medical treatment. One must be careful not to touch them either in the water or on the beach. Although their float makes a pop when stepped on, it is not recommended to do so, even with shoes.

*Figure 5.4. (a) The ghost shrimp (*Callianassa major*) is a swash zone (b) burrower and may go deep, pumping its fecal pellets to the surface.*

*Figure 5.5. A Portuguese man-of-war (*Physalia*) that has washed up on a Texas beach.*

Backbeach

The dry backbeach is a very important habitat for a few species of animals. The most notable on the Gulf Coast is the sea turtle community (figure 5.6). The dry beach is where the turtles crawl ashore (figure 5.7), make their nest, and lay eggs (figure 5.8). There must be a good width of dry beach or the turtles will not cross over to the landward side to make nests. This activity takes place in the dark with most species, so by morning there will be evidence of the nest location. On developed shore areas turtles may resist this landward excursion because of the many lights. Most of the coastal businesses in Florida are required to turn off night lights during turtle nesting season.

The turtles sculpt the surface over their eggs to make the nest difficult to recognize. It is important that these nests are undisturbed for several weeks while the eggs incubate (figure 5.9). Most all local governments or coastal parks have daily searches by someone on an all-terrain vehicle to find new nests. Once located, they are typically surrounded by stakes and marking tape to prevent disturbance during incubation (figure 5.10).

There are some potential problems with the eggs in that the sex of the hatchlings is linked to the temperature of the sand during the incubation period. If it reaches more than 90 °F, there is a tendency for all eggs to hatch as females. Obviously, this is not a healthy situation for the survival of the turtle population. This occurs on the Florida coast, where most of the nests are left to natural processes until hatching takes place. By contrast, the nests on the Texas coast are harvested by people trained in the handling of the eggs. They are then transferred to

▲ Figure 5.6. A Kemp's Ridley turtle making its way across the foreshore on the Texas coast. Photo courtesy Padre Island National Seashore.

▶ Figure 5.7. Track of a turtle as it appears on the dry beach. This is generally the best way to recognize the location of a nest. Photo courtesy National Park Service.

incubators, where they are kept under strict environmental control until the hatch. The hatchlings are then released on the beach, generally in front of a good-sized crowd of interested folks (figure 5.11).

The other quite popular beach animal is the ghost crab (*Ocypoda quadrata*; figure 5.12). These crabs are quite mobile, breathe oxygen, and are excellent burrowers. Their quickness is rather amazing as they typically move rapidly sideways across the backbeach. Generally these animals are nocturnal, thus the name ghost crab. They are very rarely seen during daylight hours, but at night they seem to be everywhere. The burrow of these crabs is rather large in diameter at the surface and has excavated sand adjacent to the surface opening (figure 5.13).

Figure 5.8. Turtle making a nest and laying its eggs on the backbeach. Photo courtesy National Park Service.

Figure 5.9. Several turtle eggs as they appear in a nest. Each egg is about 3.8 cm in diameter. Photo courtesy Padre Island National Seashore.

Figure 5.10. A marked turtle nest along the Florida coast.

Figure 5.11. Hatchlings of sea turtles on the beach at Padre Island National Seashore. The individuals are about 5 cm long. Photo courtesy National Park Service.

A few species of plants are common on the backbeach. They are on the most distant part of the beach from the shoreline. The beach morning glory (*Ipomoea*) or railroad vine is quite common and covers much of the area (figure 5.14). This species is very opportunistic and tends to be the first vegetation that covers the landward portion of the dry beach. These vines need little

COMMON ANIMALS AND PLANTS 99

Figure 5.12. A ghost crab (Ocypoda quadrata) on the backbeach. Photo courtesy Gay Hejtmancik.

Figure 5.13. Burrow and excavation sand at a ghost crab burrow on the dry beach. Photo courtesy Rebecca in the Woods.

Figure 5.14. Beach morning glory (Ipomoea) on the backbeach, where it extends its vines many meters and puts down roots along the way. It comes in (a) lavender and (b) white. Photo a, courtesy Gay Hejtmancik; b, courtesy J. W. Tunnell.

moisture and may grow many inches each day. They are effective in trapping wind-blown sand in the backbeach. Eventually this accumulation of sediment becomes coppice mounds, the earliest stage of dune development. As time passes, these small sand accumulations develop into a foreshore dune ridge, which is the primary dune system landward of the beach environment.

Probably the most important plant for accumulating sand and helping to develop and preserve dunes is sea oats (*Uniola paniculata*). This protected species actually looks like oats and is quite pretty as it blows in the coastal wind (figure 5.15). The plant extends from the coppice mounds on the landward end of the beach up to the foredune complex next to the beach.

Flotsam

There are two floating colonies of organisms that have a negative impact on the beach and nearshore. These both can be found throughout the Gulf but are sporadic in their occurrence. One is called "red tide," a bloom of a dinoflagellate type of algae (*Karenia brevis*). These small organisms can occur up to hundreds per milliliter of water, and they turn the water a grayish to reddish color (figure 5.16a). This dinoflagellate contains a neurotoxin that is taken into fish gills, causing paralysis, so the fish die, commonly to wash up on the beach in huge numbers (figure 5.16b). These algae also are taken in to shellfish (oysters, shrimp, etc.), thus causing humans to become sick because cooking does not kill the toxins. Their presence in the air can also cause human respiratory

Figure 5.15. Sea oats (Uniola paniculata), one of the most widespread of foredune plants. They are protected from human interference.

problems. Red tides have been known in the Gulf since the days of the Spanish explorers. It is still not known what causes these blooms, and scientists have been working on the problem for decades. They are most common in Florida in spring but occur in Texas generally in the fall.

Another floating organism that makes its way to the beach in huge volumes is the brown algae *Sargassum*. This macroalgae was first discovered in the middle of the Atlantic where it floated in huge colonies. It is called *Sargassum* from the Greek word for "grape" because the structure is complex with "leaves" and small spherical floats that resemble grapes. The early explorers of the central North Atlantic called the area the Sargasso Sea. The algae are also quite widespread in the Gulf of Mexico and respond readily to wind. It is common for a storm to blow huge volumes onto the beach in rows, especially on the Texas coast. The colonies of the algae contain a small community of organisms that "travel" with it for both food and protection (figure 5.17).

SUGGESTED READING

Ingle, L., and H. S. Zim. 1989. *Seashore Life: A Golden Guide*. New York: St. Martin's Press.
Witherington, B., and D. Witherington. 2007. *Florida's Living Beaches: A Guide to the Serious Beachcomber*. Sarasota, FL: Pineapple Press.

Figure 5.16. Red tide (Karenia brevis), a dinoflagellate algae that causes problems for the beach and surf zone area: (a) as seen from satellite on the Florida coast and (b) dead fish at the shoreline on the Texas coast. Photo a, courtesy NASA.

Figure 5.17. (a) Large accumulations of Sargassum, a floating brown algae that houses (b) a small community of organisms as it moves about the open water. Storms bring large quantities to the beach and disrupt recreation with the smell. Photo a, courtesy NOAA; photo b, courtesy Kim Withers and Tony Amos.

PART

BEACHES ALONG THE GULF OF MEXICO COAST

The discussion of the beaches around the Gulf covers specific geographic areas: individual states in the United States, and Mexico and Cuba combined. Even though beaches have wide-ranging characteristics on a global basis, those of the Gulf of Mexico have many common features. With few exceptions the beaches are on barrier islands that surround the Gulf. All are exposed to microtidal conditions (tidal range <2 m), and all have low to moderate wave energy because they are fetch limited. Hurricanes can impact, and have impacted, the entire coast. Except along the northern coast of Cuba and a small reach of Mexico, the continental shelf is rather wide and gently sloping throughout. The two biggest differences among beaches within the Gulf are the dominant composition of the beach sediment and the existence of rocky, high-relief shores. Virtually all of the beaches of the United States are dominated by terrigenous sand except the Florida Keys, whereas significant parts of Mexico and Cuba are carbonate sediments. Rocky shores are essentially absent in the United States, but Mexico and Cuba have a broad combination of sand and rocky shore environments.

Florida has one of the most human-developed coasts and diverse beach systems of all of the Gulf Coast. The southernmost part is dominated by carbonate sand and gravel, and extensive coastal reaches do not have any wide beaches. Barrier islands dominate the Gulf peninsula and the Panhandle. Both Alabama and Mississippi have only short coasts that are dominated by barrier islands with good beaches and some intensive development. Louisiana is quite different because it is dominated by the Mississippi River delta complex and the associated, fragile barrier islands. Probably the most uniform beach system is that of the Texas coast. It consists of nearly all wave-dominated barriers with wide beaches. Only three portions of the Texas coast are heavily developed for human activity: in the north, the Galveston to Freeport region; in the middle, near Corpus Christi; and in the south, at South Padre Island. There are extensive coastal reaches between these developed sections that are essentially pristine, with each including many tens of kilometers of beach.

The discussion of the beaches of Mexico and Cuba is limited because there is much less known about them than US beaches. Many of the coastal areas of both of these countries are inaccessible, and only a few population centers are present on these coasts. Mexico's beaches range widely from rocky beaches of limestone and volcanic material to carbonate sand and gravel. Human development is modest with the medium-sized cities of Veracruz and Tampico, along with the intense development at Cancún. Cuba also contains a range of beach morphology and rocky coasts. Both terrigenous and carbonate beaches are present.

The popularity of the beach as a tourist destination coupled with widespread erosion problems has led to beach nourishment to correct erosion problems throughout many parts of the Gulf Coast. A range of approaches has been implemented to address this problem. Some of these projects have been very successful, such as along the Florida peninsular coast, and others have not fared so well, such as on Galveston Island in Texas.

The treatment of beaches geographically begins with the eastern part of the Florida Keys and proceeds anticlockwise around the Gulf, ending with the Varadero Beach area of Cuba. One should remember that the beach is a very dynamic environment and changes are common, sometimes very significant, especially after storms. The examples shown in the following chapters might not be just like the beach appears at the time you visit. Most of the beach photos were taken within a week in 2012. The photos of the coast of Mexico are courtesy of J. W. Tunnell and were mostly taken in the 1980s.

6

Beaches of Florida

FLORIDA beaches can be easily subdivided into three distinct regions: the Keys, the Gulf peninsula, and the Panhandle. Each of these regions has its own characteristics and its own types of beaches. The three sections are also each oriented quite differently to the primary weather patterns, they have different offshore regions, and they experience hurricanes differently. It should also be noted that these regions are separated by extensive coastal reaches where beaches are rare and poorly developed: the Ten Thousand Islands mangrove system and the Big Bend coastal marsh system (figure 6.1).

Florida Keys

Although not a part of the Florida Keys, the islands of the Dry Tortugas are a part of Florida and have beaches. These islands are associated with extensive reef development on Quaternary carbonates and are occupied only by Fort Jefferson, a national monument, and a Coast Guard station. The beaches there are accumulations of reef debris that is coarse sand and all calcium carbonate (figure 6.2).

The Florida Keys might be considered to be on the Atlantic side of the state but are here treated as part of the Gulf of Mexico, as is the Cuban coast across from them. The Keys themselves represent carbonate accumulations that were deposited about 120,000 years ago when sea level was about 3–5 m above its present position. The Upper Keys are Key Largo Limestone, which is essentially Quaternary coral reefs, and the Lower Keys are Miami Oolite, a limestone formation that originated as tidally influenced shoals of spherical carbonate sand grains (figure 6.3). The Key Largo Limestone produces a rather linear coastline

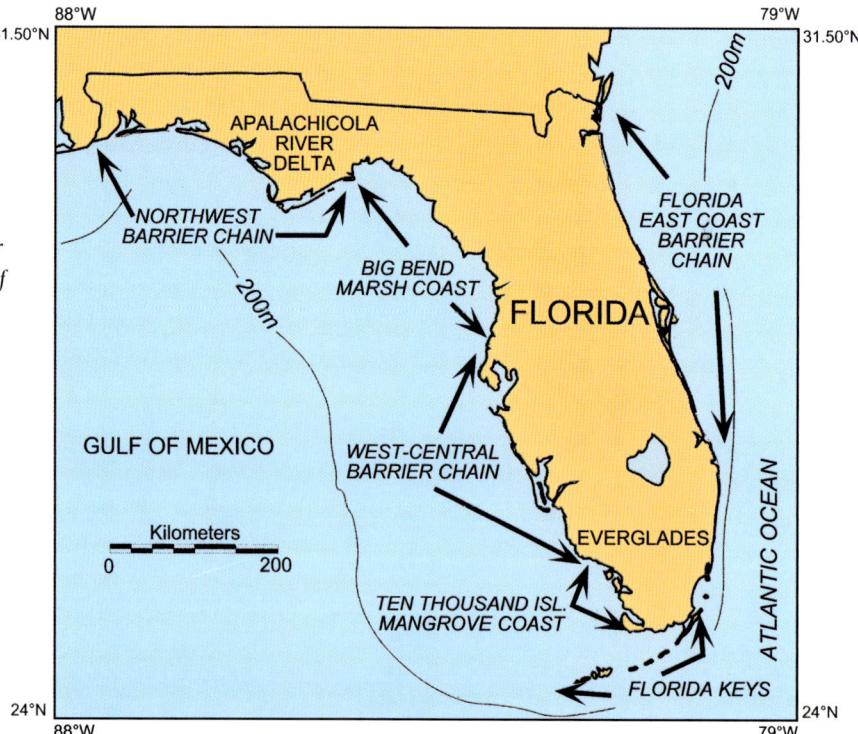

Figure 6.1. Outline map of Florida indicating the major coastal provinces of the Gulf Coast.

Figure 6.2. Beach of carbonate sediment derived from reef debris on East Key of the Dry Tortugas, located about 50 km west of Key West.

overall, and the system is narrow. By contrast, the Miami Oolite formation displays a wide geography with linear trends at right angles to the Upper Keys of the Key Largo reefs. Basically, the Key Largo reefs were wave dominated, and the Miami Oolite sand bodies were tide dominated. The beaches associated with these Keys or islands are carbonate sand and gravel, virtually all of which is biogenic. There is a significant exception in a nourished beach in Key West.

In general, the beaches on the Keys are not wide, and few are good for recreation. In some places the shoreline is rocky limestone or beach material is

Figure 6.3. A satellite image of the Florida Keys showing the difference between the Upper Keys (Key Largo Limestone) and the Lower Keys (Miami Oolite). Image from NASA.

scattered. Limestone does not produce good beach sand material because it does not break up into that form or particle size. The local environments are producing some skeletal material but not enough to produce good beaches.

The Keys are quite vulnerable to tropical storms and hurricanes. Their elevation is typically less than 2 or 3 m above sea level, so storm surges can be devastating to buildings and infrastructure. Vehicular accessibility is by one road only, making evacuation very difficult in the event of a storm. Numerous hurricanes have impacted these islands over the past several decades and caused tremendous damage. These events are really a problem for maintaining good recreational beaches.

An example of a poorly developed beach on the Key Largo Limestone is Anne's Beach on Lower Matecumbe Key (figure 6.4). Here the Key Largo Limestone crops out at the shoreline, but there is a small accumulation of carbonate sand that is typical beach material. It extends only a few meters along the shoreline but is enough for a few people to use.

The best natural beaches in the Keys are those at Bahia Honda State Park where three beaches are present (figure 6.5). This is the most popular park in the Keys largely because of the beaches. Smathers Beach in Key West is the main beach in this area (figure 6.6). It is unusual in that it is predominantly quartz sand instead of carbonate, because it was nourished multiple times with both quartz sand that was trucked to the site and carbonate, ooid sand that came by boat from the Bahamas. In both cases the transportation to this location made beach nourishment quite expensive.

Figure 6.4. Anne's Beach on Matacumbe Key is very small in both width and length, but it is popular because the choices in the area are limited.

Figure 6.5. Bahia Honda State Park in the Keys with abundant Sargassum *on the active beach.*

Figure 6.6. Smathers Beach in Key West is very popular because of its location. It has been nourished with both upland terrigenous sand and carbonate ooid sand from the Bahamas.

Gulf Peninsula Beaches

The southwestern Florida coast is dominated by mangroves and tidal creeks. Waves are small and sediment is scarce. The only semblance of a beach is in the form of shell concentrations along some of the mangrove mangal shorelines (figure 6.7).

The Gulf Coast of the Florida peninsula is probably the most diverse barrier-inlet system in the world. It includes 30 inlets and a similar number of barrier islands (figure 6.8). Each of the barriers has a well-developed beach, although erosion is common on many of them. Public access is widespread along this entire coast both at designated parks and at the ends of many roads and streets where parking is available. In a few locations parking is free, but most require payment. Beach nourishment has taken place at several locations, and sometimes multiple times. Human development is widespread except at parks or on those islands not accessible by road.

Figure 6.7. Mangrove mangal island in southwestern Florida with a small, shelly beach.

Many of the beaches of the Florida Gulf Coast have been nourished to provide continuing recreational venues for tourists. In general, this coast is sediment starved, which affects the beaches themselves because storms can remove a lot of sediment. Some is carried offshore, but much of that can make its way back to the beach. A lot of sand is carried alongshore by currents during a storm. This sand commonly comes to rest on the ebb-tidal deltas at tidal inlets. The source of nourishment sand comes from the ebb-tidal deltas and offshore sandbars. The distance of travel from the borrow site to the construction site is a major factor in the cost per unit for the material.

Such projects are very expensive and are funded in different ways, including private and various levels of government sources. The State of Florida has designated $30 million per year to fund its share of the costs. Each project ranges up to multiple kilometers in length and several hundred thousand to more than a million cubic meters of borrow material. Most of the projects on this coast have been successful, but renourishment after five to seven years, or sometimes less, is common.

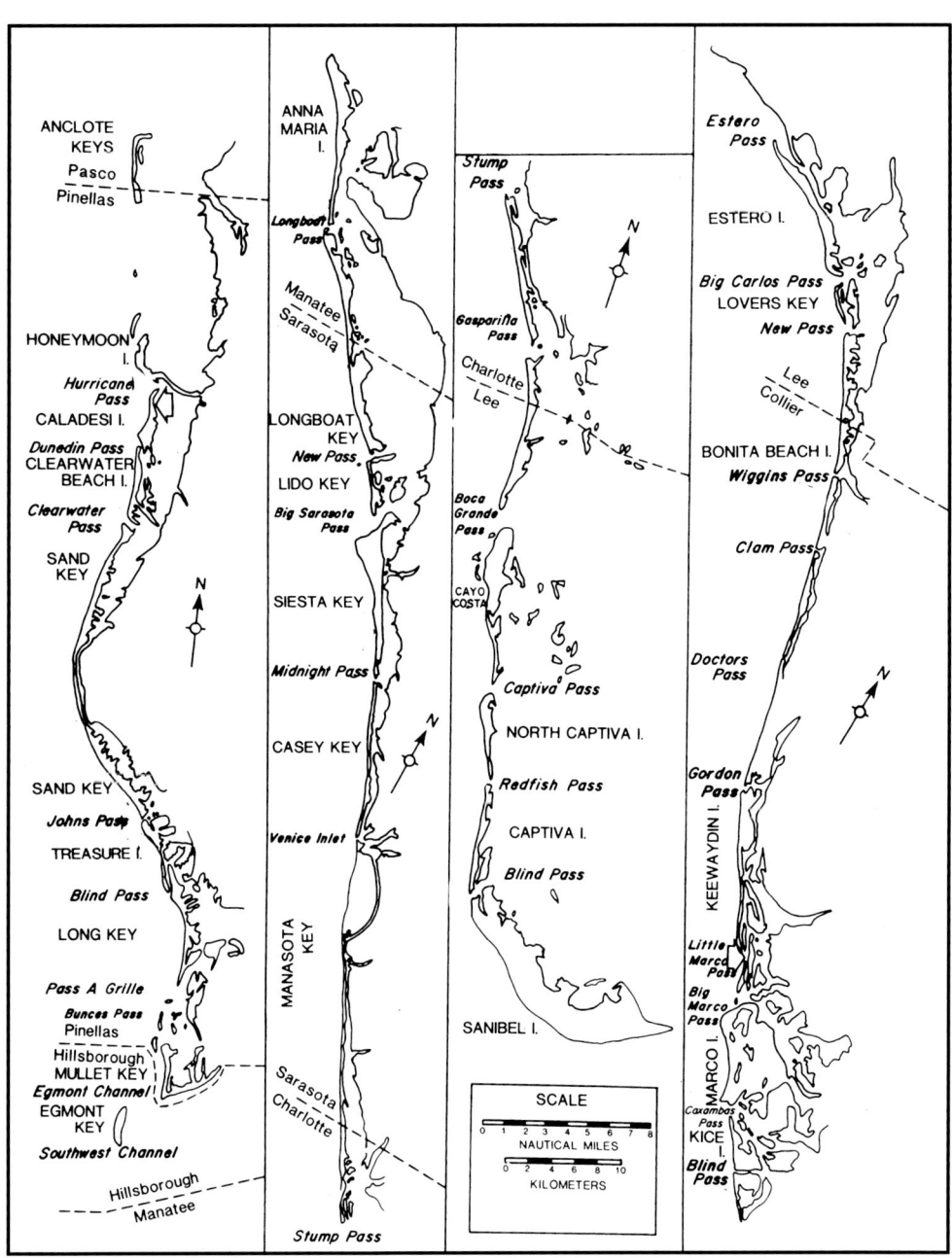

Figure 6.8. Map of the barrier-inlet system on the Florida Gulf Coast, probably the most complex barrier system in the world. Modified from R. A. Davis, "Morphodynamics of the West-Central Florida Barrier System: The Delicate Balance between Wave and Tide Domination," in Coastal Lowlands: Geology and Geotechnology *(Dordrecht, Netherlands: Kluwer, 1988), 225–35.*

The most pristine barrier islands include the southernmost one, Keewaydin Island, the State Parks of Caladesi and Honeymoon Islands, and Anclote Key. North Captiva and Cayo Costa Islands have no causeways, but only a few residences are present on each. These barriers have very nice beaches, and erosion is a problem only on the central portion of Honeymoon Island, where rocky fill was placed in the early 1960s. The sediment on these beaches is dominated by terrigenous sand, mostly quartz with varying amounts of shell and shell debris.

Marco Island / Naples

These two coastal regions are among the wealthiest on the Florida coast. High-rise condominiums line the beaches, especially on Marco. The wide beaches are very beautiful, with blue-green water in the modest surf (figure 6.9). There is a major contrast between the two areas regarding public access to the beaches: virtually none on Marco, but there are numerous places for public access in Naples, most with parking available (figure 6.10). The areas north of Naples, such as Vanderbilt Beach, Bonita Beach, and Fort Myers Beach, all have excellent beaches for recreation.

Figure 6.9. Nourished beach on the southern portion of Marco Island.

Figure 6.10. An excellent example of a Naples beach adjacent to the downtown area.

Sanibel and Captiva Islands

One of the best-known beaches on the entire Gulf is at Sanibel Island. It is probably the most accessible, good shell-collecting location anywhere in the United States. Near the southern end of the barrier, the beach is essentially all shell material, much of which is whole shells. Both of these islands have excellent beaches. Those at Sanibel tend to be wide and gently sloping, whereas beaches on Captiva are steeper and narrower.

Multiple hurricanes in the first few years of this century have produced considerable erosion on these islands, which resulted in the need for extensive nourishment. Sanibel Island is typically a sediment sink at the end of a north-to-south moving longshore transport drift system. It has a wide beach, and

Figure 6.11. The beach at Sanibel Island is quite shelly, wide, and gently sloping. The island "turns a corner" at its southern end. Sanibel beaches vary in appearance, including (a) wide and natural profiles, (b) vegetated backbeach, (c) areas of planted grasses, (d) coppice mounds, and (e) abundant shells (opposite).

various efforts have been made to stabilize it and build dunes. Shells may be locally concentrated to dominate beach sediment (figure 6.11).

Captiva Island contrasts with Sanibel in that it is narrow, erosional, and developed mostly by large, single-family homes with a few resorts. A section of the island has been eroded back to the paved road and was subsequently armored with riprap and then nourished (figure 6.12). Farther to the north the nourished beach is wider and less steep in a major tourist area (figure 6.13).

North Captiva Island and Cayo Costa Island are next in the barrier-inlet system and are quite different from each other. Neither of these islands is accessible by vehicle, so they are not significantly developed. North Captiva is narrow with narrow beaches and is erosional at its southern end. Cayo Costa is a wide barrier with multiple sets of prograding beach ridges and wide beaches (figure 6.14). It has the most complicated geomorphology of all the barriers on the entire Florida coast. In addition to the beach ridge sets, there are offshore, supratidal shoals.

Gasparilla Island / Don Pedro Island / Manasota Key

The beach systems along this reach of the Florida Gulf Coast are essentially natural; nourishment is not a factor in their condition. They are very popular residential and tourist destinations but are dominated by single-family homes and small resorts. In general, the beaches are in pretty good shape. Small public parks are common and popular (figure 6.15).

The beaches at Manasota Key show significant accretion (figure 6.16). The wide, vegetated backbeach area suggests that all of that portion of the beach has accumulated in very recent time, a situation common along this part of the coast where barrier islands are wave dominated.

Figure 6.12. The Captiva Island beach has been nourished multiple times. It is narrower and steeper than that at Sanibel to the south. This is the steepest location shown (a) before nourishment in the 1970s and (b) after in 2012.

Figure 6.13. The wide part of the nourished part of Captiva Island at the South Seas Plantation resort.

Figure 6.14. Aerial view showing the complicated nature of Cayo Costa Island with multiple sets of prograding beach ridges and supratidal offshore shoals. Scale bar is 1 km.

Figure 6.15. (a) The south end of Gasparilla Island showing a well-developed beach. The heavy line points to a concrete seawall that was constructed in the 1950s when erosion here was a problem. (b) Wide natural beach at a small park on Gasparilla Island.

Venice Area

The only portion of the Florida peninsula coast that is not fronted by barrier islands is just south of the small city of Venice where the Miocene mainland strata extend to the shoreline and beyond. In this area the Intracoastal Waterway (ICW) had to be cut and dredged through the mainland to connect the back-barrier bays that typify this coast. The beach in the southern part of the area is in pretty good shape (figure 6.17a), but the nature of the beach sediment is noticeably different from that of the coast overall. The dark-colored sand is phosphate material that is reworked from the Miocene deposits that are

Figure 6.16. Significant accretion on this wide beach has taken place in a rather short time as evidenced by the grass in the backbeach environment.

Figure 6.17. Public beach south of Venice on the mainland: (a) wide beach and (b) close-up showing the dark-colored sand due to the presence of abundant phosphate from the Miocene strata.

present along this part of the coast (figure 6.17b). Major areas in central Florida are mined for this phosphate. Farther to the north in the Venice area, the beach is erosional and nourishment has taken place (figure 6.18).

Across the Venice Inlet in Nokomis at the southern end of Casey Key is a well-known beach where tourists spend much time and effort in search of shark's teeth (figure 6.19a). These shark's teeth are found in some abundance along with skate and ray teeth, bone fragments, and other phosphatic remains of vertebrates from the Miocene strata (figure 6.19b). Most of the teeth are rather small, less than 1 cm wide, but some have been found that are more than 5 cm across.

Figure 6.18. Severe erosion forming a distinct scarp on a recently nourished beach in the town of Venice.

Figure 6.19. (a) Public beach just north of the Venice Inlet in the background. (b) Close-up of shark's teeth from the surf zone area offshore of this beach. Scale bar is 1 cm.

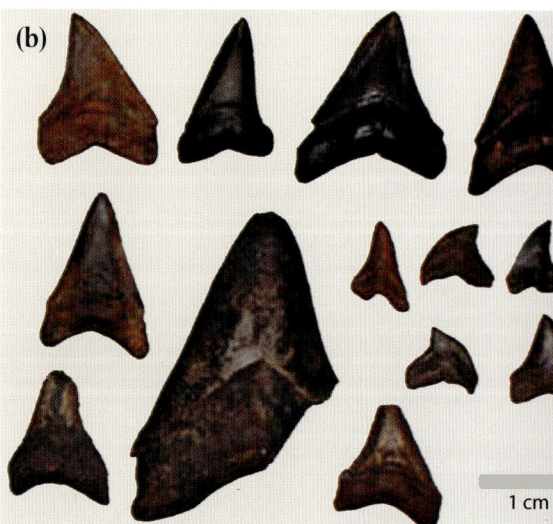

Figure 6.20. The public beach at Turtle Beach on Siesta Key. The abundance of footprints testifies to its popularity.

Casey and Siesta Keys

In some respects Casey and Siesta Keys are not one long barrier island. In the mid-1980s Midnight Pass, which had separated them for more than a century, was closed by natural causes—too much south-to-north longshore sediment transport. There have been many attempts to open it, but it has not happened to this date. As a result, the system is operating like one very long barrier: wave dominated over the southern two-thirds of its length and mixed energy on the northern part of Siesta Key. The Casey Key portion is very narrow with a narrow and rather steep beach. It is completely private with no public access.

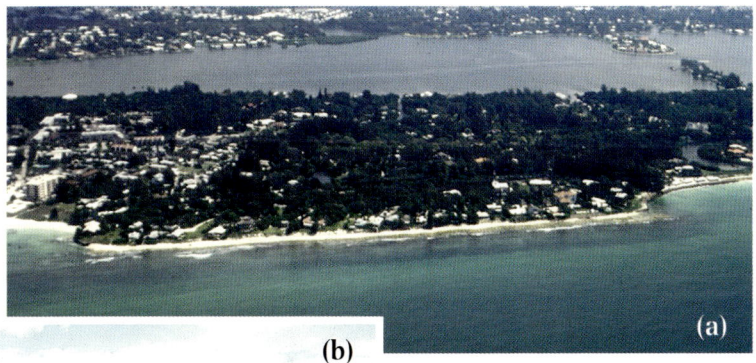

Figure 6.21. Point of Rocks is the only US Gulf Coast rocky coast: (a) aerial view showing the extent and erosion resistance of Point of Rocks and (b) close-up showing the nature of the beach strata of the rocky coast.

Figure 6.22. An oblique aerial view showing how sand is bypassing Big Sarasota Pass and attaching to the public beach area on Siesta Key.

Recall that without public funds for nourishment, there is no requirement for public access to the beach.

Siesta Key is very wide on its northern end but is also narrow for a few kilometers north of Midnight Pass. Turtle Beach is one of the good public beaches available (figure 6.20). Between there and the northern end of the island is a unique part of the Gulf Coast, Point of Rocks (figure 6.21). This is a true rocky coast, the only one on the US Gulf of Mexico. This short reach of coast is a type of beachrock in that it was deposited as beach sediment that has subsequently become lithified.

The public beach near the northern end of Siesta Key is one of the highest-rated beaches in the United States. This beach is very wide and is continually adding beautiful sand (figure 6.22). The only problem is that it is so attractive that it is generally very crowded and parking is at a premium.

Lido and Longboat Keys / Anna Maria Island

The next three barriers, Lido and Longboat Keys and Anna Maria Island, have all been nourished and also have hard structures, mostly in the form of groins. They are densely developed, especially Longboat Key.

Lido Key has been nourished regularly and serves as a source of sediment for the beach at Siesta Key. The north-to-south longshore sediment transport

Figure 6.23. Recently nourished beach on Longboat Key as it appeared in 1996.

Figure 6.24. Longboat Key beach was nourished in 1996 and has performed quite well as evidenced by its width and the stabilizing vegetation through the backbeach

▲ *Figure 6.25. Southern Anna Maria Island showing a wide beach with groins in the background.*

▲▶ *Figure 6.26. Narrow, steep beach in the northern part of Anna Maria Island.*

▶ *Figure 6.27. Oblique aerial view of Egmont Key showing expanding beach in the foreground (south) and erosional conditions in the north.*

carries sediment across the mouth of Big Sarasota Pass and eventually welds to the beach at Siesta. Longboat Key is an upscale community that has had significant erosion problems. Nourishment has been helpful, but there are significant hot spots where the rate of erosion is quite high and that require regular attention and local nourishment. A major nourishment project took place in 1996 (figure 6.23). Several years later the beach on part of the island was still wide with much of the backbeach being stabilized by vegetation (figure 6.24), but other areas needed renourishment.

Figure 6.28. North Bunces Key, a very young island with a well-developed beach.

Figure 6.29. Nourished beach on Pass-A-Grille, one of the most popular beaches in this part of the Florida Gulf Coast.

Anna Maria Island has several groins at its southern end that are part of the numerous groin fields constructed along the central Florida Gulf peninsula in the 1950s when erosion was widespread and severe (figure 6.25). Nourishment has been concentrated toward the northern half of the island and has been only moderately successful. Erosion is a serious problem, and the nourished beaches become narrow and steep in a short time (figure 6.26). This island mimics Siesta Key in general morphology but does not have the luxury of a sediment supply from the north.

Figure 6.30. An example of a poor beach management practice that is fairly common on this part of the coast. All vegetation has been removed in front of this small tourist hotel, making the beach susceptible to erosion.

Figure 6.31. Before and after photos of the Upham Beach area showing the difference that nourishment makes and the duration of the improvement. Photos courtesy Nicole Elko.

Egmont Key / Fort De Soto Park / North Bunces Key

The mouth of Tampa Bay has a complex of small islands with popular beaches that are essentially pristine. Egmont Key is a small island directly in the mouth of Tampa Bay that is part of the huge ebb-tidal delta. It had military installations associated with the Spanish-American War and is home to the harbor pilots for the Port of Tampa. The beach is erosional in most places and accretional at the southern end, which is a bird sanctuary (figure 6.27). It is a great place to visit by boat for a day at the beach.

Fort De Soto Park occupies a right-angle barrier island that has beaches bordering both the open Gulf of Mexico and the entrance to Tampa Bay. Although it is an excellent recreational area and is very popular, because the park is so large, it does not get crowded. The gulfside beach is expanding as sediment is carried landward across the shallow ebb delta. Across Bunces Pass is North Bunces Key, a barrier island that originated from wave-generated sand shoals in the 1960s. This beach is accreting and dunes are forming (figure 6.28). Because of its relatively remote situation, it is a major site for birds.

Figure 6.32. Post-nourishment condition at Upham Beach, one of the most erosional beaches on the Gulf.

Long Key / Treasure Island

Most of the beaches in this coastal reach have been nourished, several multiple times (figure 6.29). The result of this practice is abundant locations to access the beach, some with parking. Even though nourishment is providing excellent beaches for tourists, there needs to be some education for the business owners and their cooperation to maintain these beaches. Practices such as removing all of the vegetation to provide a total beach across the entire coastal property is poor beach management (figure 6.30). Lack of any stabilizing factors such as plants will make beaches quite vulnerable to erosion and require more frequent, high-cost renourishment.

At the northern end of Long Key is one of the most popular beaches, Upham Beach. This location is just downdrift of Blind Pass, a small tidal inlet that separates Treasure Island from Long Key. As a result of its location, Upham Beach is probably the most critically eroded portion of this coast, and in the past it was nourished every second or third year (figure 6.31). More recently, geotextile tube groins were added to help maintain the nourished beach; however, erosion continues (figure 6.32).

Treasure Island to the north is also nourished near its middle portion because of severe erosion problems (figure 6.33). Erosion is fairly widespread,

▲ Figure 6.33. Beach erosion and buildings in jeopardy on part of Long Key, which is the rationale for nourishment in this area.

▲ ▶ Figure 6.34. Probably the widest beach on the Gulf Coast near the north end of Treasure Island. It would be better to permit vegetation to stabilize the beach instead of regular raking, which removes it.

and some of the buildings are threatened in the absence of nourishment. By contrast, the northern portion of this island has what is probably the widest beach on this coast because of the dredge spoil from Johns Pass to the north that was placed there in the early 1970s (figure 6.34). Unfortunately, beach management practices here by hotels do not permit vegetation, and therefore dunes are not being nurtured. Their presence would enhance the stability of this location.

Sand Key

Sand Key is the longest barrier island on the Florida coast and has numerous public beaches as well as abundant parking and access. In the past it has been the location of many shore-protection structures, including extensive seawalls, groin fields, a breakwater, and other attempts to maintain a beach. Beach nourishment along this section of coast has since become the primary method for beach protection and has been very successful, more so than in nearly all other parts of the Florida coast. Although nourishment must be redone periodically, the cost-benefit ratio is very positive because it is a densely populated area that has high tourist visitation.

The beaches along this part of Florida have a high rate of turtle nesting for the Gulf of Mexico. There are, however, multiple factors that cause problems for nesting turtles, including the abundance of tourists, lights on the beach at night, and hard-packed shelly sand in some locations. No beach nourishment is permitted during nesting season, from May through October, because of the obvious problems for nesting or for existing nests. Additionally, the state requires that all beaches in this region be cultivated prior to the nesting season to be certain that the sand is soft enough for turtles to dig their nest sites. The entire county is traversed early each day to locate fresh nests, and then each nest must be fenced off to protect it. Because of the intense development along the beach, all lighting on the beach must be turned off because it inhibits turtles from nesting. On this coast the nests are left undisturbed and well marked, so hatchings and dispersal to the Gulf are natural.

Sand Key fronts a bedrock headland area on this coast, and the island is only about 100 m wide at its narrowest part. This is also where the longshore sediment transport diverges to the south and the north. The northern end of the barrier has become one of the most popular and widest beaches since construction of the jetty at Clearwater Pass was completed in 1976. This beach is fed by the nourishment near the headland on Sand Key, from which sand is carried to the north toward Clearwater Pass (figure 6.35). Prior to the construction of the jetty, this sand was carried into the tidal inlet, causing navigational problems.

Figure 6.35. Wide beach in the fillet on the northern end of Sand Key at the jetty at Clearwater Pass.

Figure 6.36. Beautiful, wide beach on the nontourist portion of Clearwater Beach Island.

Figure 6.37. Aerial view of the northern portion of Clearwater Beach Island showing the amount of beach progradation in front of the seawall built in the 1950s (black lines).

Clearwater Beach and Caladesi Island

Clearwater Beach is probably one of the most famous tourist destinations on the Gulf Coast (figure 6.36). This barrier island is completely developed with excellent beaches. Unlike most of the coast south of here for many kilometers, this barrier has not been nourished. In the early 1950s there was a severe erosion problem along Clearwater Beach Island. A seawall was constructed to protect the buildings along most of the island. Since then, there has been nearly 100 m of accretion of beach and dune sand gulfward of the wall (figure 6.37). Because there has been no beach nourishment on this island, access to much of it is limited.

Figure 6.38. White sand beach on Caladesi Island with ridge and runnel at low tide. Seaweed/algae is common on the beaches of this coast after storms.

Figure 6.39. Transgressive, erosional nature of the sediment-starved northern end of Caladesi Island. The mangrove tree trunks and mangrove peats are remnants of the back-barrier environments.

Clearwater Beach Island and Caladesi Island were separated by Dunedin Pass until the tidal inlet closed in 1988. This made the two islands essentially one, much like Casey Key and Siesta Key to the south. Caladesi is a popular state park accessible by boat or by walking most of the length of Clearwater Beach. The beach at Caladesi has been rated among the top 10 in the United States (figure 6.38). At times the northern end of the island shows the typical erosional, transgressive nature of a drumstick barrier (figure 6.39). Mangrove trunks and marsh peat are exposed as the small amount of sand at this end of the island is carried landward.

Honeymoon Island / Three-Rooker Bar / Anclote Key

The northernmost three islands of this coast are all undeveloped and publicly owned. Honeymoon Island is a state park, the second most popular in Florida. It has had erosion problems since the failed attempt at development in the early 1960s. Nourishment has been necessary on the southern portion where people visit (figure 6.40), and a terminal groin has been added to help stabilize the sand. The spit that forms the northern end of the island actually has the best beach. Sediment transport diverges at the change in direction of the shoreline, and much sand is carried north; it is a long walk from the parking lot, however.

Three-Rooker Bar is a small, natural barrier that has developed from sand shoals since the early 1970s (figure 6.41). It is a favorite day recreation area for boaters. The northernmost barrier on this coast is Anclote Key, a narrow, wave-dominated barrier with good beaches throughout (figure 6.42). This island is a favorite place for weekend campers, but there are no facilities.

There are no natural beaches of any consequence north of Anclote Key (figure 6.43). This section of the coast is the Big Bend area and is dominated by wetlands along the open Gulf shoreline. There are some small public sites that provide a beach experience, mostly visited by locals. These include Green

Figure 6.40. Aerial overview of Honeymoon Island with beach accretion in the foreground and on the northern spit of the island in the upper right.

Figure 6.41. Oblique view of northern Three-Rooker Bar between Honeymoon Island and Anclote Key, showing sand accretion on the beach.

Figure 6.42. Anclote Key is a wave-dominated, narrow barrier with well-developed beaches that are growing with additional sand.

Key and Hernando Beach in Pasco County and Crystal River Beach in Citrus County (figures 6.44 and 6.45). These popular facilities were constructed by local governmental agencies to provide residents the opportunity of a nearby beach experience.

Figure 6.43. Oblique view of the Big Bend coast showing its irregular shoreline and absence of beaches.

Figure 6.45. Crystal River Beach on the marsh-dominated coast, a built facility. Photo courtesy Google Earth.

Figure 6.44. Hernando Beach, a constructed facility for the local residents. Photo courtesy Google Earth.

Florida Panhandle

The Panhandle coast of Florida and its contained beaches are more like those of Alabama and Mississippi than of the Florida peninsula, primarily because of the origin of the coastal sediments and the orientation of the coast within the Gulf of Mexico. The beaches on this coast extend from Alligator Point on the east to Perdido Key at the Alabama line. This reach of wave-dominated barrier islands contains what are arguably the most beautiful beaches on the

Gulf of Mexico. Some of the coast is pristine, and some is very heavily developed for tourism. There is pretty much something for everyone.

Alligator Point is a small residential beach community that has both wide beaches and extremely erosional shorelines (figure 6.46). The beaches have developed primarily at the expense of the erosional conditions updrift. Several homes are threatened by erosion and have various protection structures in front of them (figure 6.47). These very different conditions exist within a few hundred meters of each other.

Figure 6.46. Very nice, wide accretional beach in the Alligator Point area of the Florida Panhandle.

Figure 6.47. Home on Alligator Point where erosion is extreme and protection is required.

Apalachicola Area Barrier Islands

Several barriers of quite different styles front the Apalachicola delta and bay (figure 6.48). They range from drumsticklike to distinctly wave dominated and have quite different orientations relative to wave approach. As a result, erosion and deposition vary in both space and, apparently also, in time. The most easterly is Dog Island, very natural and accessible only by plane or boat. There are about 100 inhabitants and one small, rustic hotel. It has gorgeous beaches and dunes because of its limited access and disturbance. Accretion dominates the eastern end; and erosion, the western end (figures 6.49 and 6.50).

The next island, St. George, has some natural areas and others with dense residential development. The beaches are universally good and wide. The adjacent water is clear and beautiful. There is little variation in the nature of the beach and adjacent environment throughout the island. The beach is accretional, and dunes are beginning to form. Fences have been placed in the back-beach area to help accumulate sand (figures 6.51 and 6.52).

St. Vincent Island is quite different from the others in this coastal reach. It

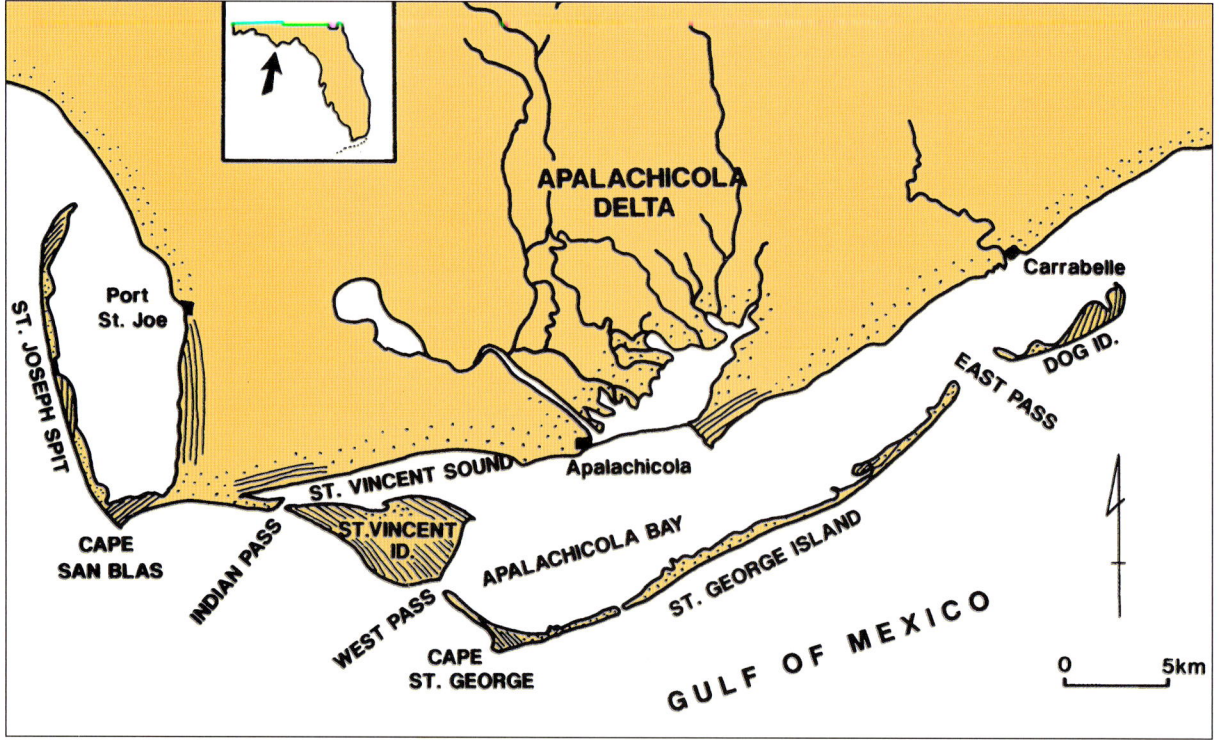

Figure 6.48. Map showing the barrier islands in the Apalachicola area.

Figure 6.49. Nice beaches on the eastern end of Dog Island where considerable sand is accumulating.

Figure 6.50. The western end of Dog Island, which is low in elevation and commonly washed over.

is drumsticklike in shape with the wide end on the east. There are more than 100 beach ridges that have apparently developed as one major set since their orientation throughout is quite similar (figure 6.53). The island is accessible only by boat, and there is no development of any kind; therefore, the beaches are pristine (figure 6.54).

Figure 6.51. The beach and adjacent area in the eastern part of St. George Island. Notice the snow fence to help accumulate sand and the small coppice mounds on the active beach. It is apparent that this is a stable, accretional beach.

Figure 6.52. On the western part of St. George Island the beach is similar to those of the entire island: wide, accretional, and small dunes being developed.

Figure 6.53. A vertical aerial view of St. Vincent Island showing the many prograding beach ridges. Photo courtesy NASA.

Figure 6.54. Oblique aerial view showing the outstanding beach and adjacent beach ridges on St. Vincent Island. Photo courtesy US Geological Survey.

BEACHES OF FLORIDA 137

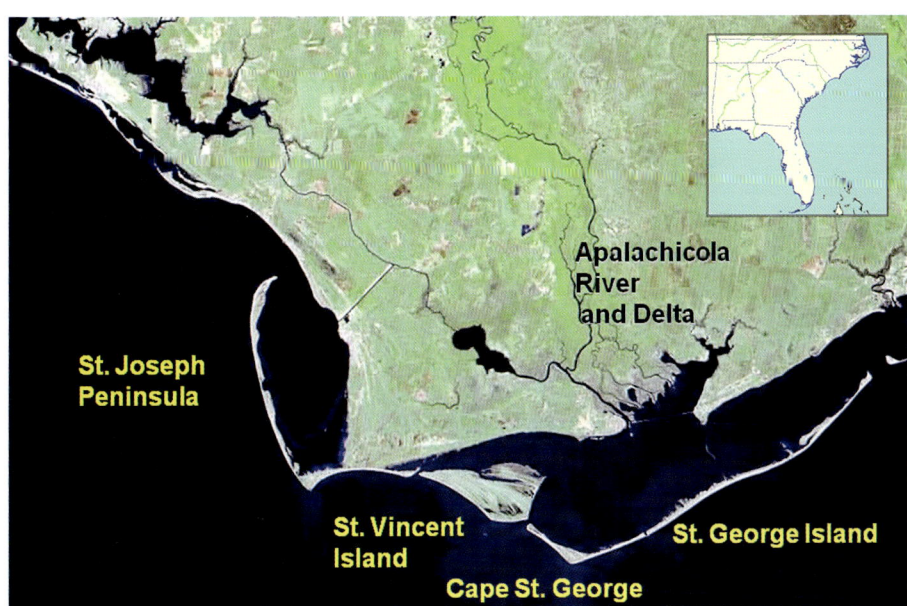

◀ *Figure 6.55. Map of the Apalachicola area showing the barrier islands and their relationship to the coast.*

▼ *Figure 6.56. Severe erosion and protection (a) on the eastern side of Cape San Blas and (b) on the western side.*

▼▼ *Figure 6.57. The beaches become wide and stable only a few hundred meters to the (a) east and the (b) west of the erosional cape apex.*

The area near Cape San Blas shows a striking contrast between very wide beaches and severe erosion, not unlike the Alligator Point area. In general, sand is eroded from the cape itself and transported to both the northeast and northwest; most goes northwest to the long spit of St. Joseph Peninsula (figure 6.55). Erosion is severe near the apex of the cape, and structures have been placed to protect the road and homes (figure 6.56). In spite of this situation, the beaches only a few hundred meters away from this area are quite wide and accretional (figure 6.57).

Panama City Area

The remainder of the Florida Panhandle is characterized by some of the best beaches on the Gulf Coast. As a consequence, there has been and is continuing to be considerable development for tourism. This area is very popular during spring break. In general the water is clear, waves are modest, and it is a wonderful region to visit the beach.

Mexico Beach is very wide, stable, and easily accessible along the coast highway (US 98). The dry backbeach includes numerous coppice mounds (figure 6.58). In addition to well-developed beaches along this coastal reach, longshore bars are nearly continuous, as evidenced by the waves that break over them.

Figure 6.58. Mexico Beach, a beautiful, wide beach with abundant coppice mounds and longshore bars that define the surf zone.

Figure 6.61. View across the beach to the surf zone west of Panama City area showing the waves breaking over a well-developed longshore bar.

Figure 6.62. Natural wide beach west of Panama City.

▲▲◀ *Figure 6.59. Beautiful, white, manicured beach in the highly developed Panama City area. These beaches attract huge numbers of tourists.*

▲▲▶ *Figure 6.60. Good example of an excellent beach with evidence of planted grasses and well-placed fences to aid in dune development.*

One of the most popular and therefore most developed sections of the coast is in the area of Panama City. It has become a major tourist destination and is also home to a very good beach system. Hurricanes have caused considerable damage, but nourishment and other mitigation approaches have resulted in little evidence of it (figure 6.59). Installing fences and planting vegetation have been generally successful (figure 6.60). Beyond the developed portion of this coastal region are vegetated dunes; a wide, stable beach; and well-developed longshore bars (figures 6.61 and 6.62).

Farther west on the Panhandle there is evidence of severe erosion of beach and dune areas from the multiple hurricanes that have struck this coast since Opal in 1995. The beach here near Sunnyside is fairly narrow and steep with high dune elevations immediately landward of the active beach (figure 6.63).

Fort Walton Beach / Pensacola Area

Figure 6.63. The beach near Sunnyside is narrow and steep with high, steep bluffs landward, testifying to intense erosion in the not-too-distant past.

This part of the Panhandle has what are the most beautiful beaches in Florida. It includes the most pristine because they are located on Eglin Air Force Base, where access is prohibited, and some of the most developed near the cities of Fort Walton Beach and Pensacola. In general the beaches are pure white quartz sand, and the water is clear. Shells are quite rare along this reach of coast.

The barrier of Santa Rosa includes a stretch of several kilometers that is off-limits because it is under military jurisdiction. The beaches here are spectacular. They are completely natural and will be as long as the status quo remains (figure 6.64). The community of Pensacola Beach and beyond to Fort Pickens State Park are areas hit hardest by recent hurricanes, especially Opal in 1995, Ivan in 2004, and Dennis in 2005. The road to Fort Pickens was completely washed out and has since been rebuilt. The elevation of the island here is very low, allowing complete washover by storm surge. Now, there is little evidence of these storms because the beaches have been naturally rebuilt and dunes are in the process of being rebuilt by natural processes (figure 6.65).

Perdido Key

The last of the Florida Panhandle barrier islands is Perdido Key, a long barrier that extends into Alabama. The beaches here have been nourished multiple times. Hurricane Ivan was the reason for a major construction project. A recent nourishment project was completed in late 2011. The result has been outstanding beaches fronting both the private developed areas and the Gulf Islands National Seashore / Perdido State Park (figure 6.66).

BEACHES OF FLORIDA 141

Figure 6.64. The beaches on Eglin Air Force Base (a-aerial view; b-ground view) are pristine, wide, and very nice. Photo courtesy R. Kirby.

Figure 6.65. Beaches on (a) the developed part of Santa Rosa Island and in (b) the Fort Pickens area. Photo b, courtesy R. Kirby.

▶ *Figure 6.66. Extremely wide beach as the result of nourishment on the park at Perdido Key.*

SUGGESTED READING

Brooks, G. R., L. J. Doyle, R. A. Davis, N. T. DeWitt, and B. G. Suthard. "Patterns and Contents of Surface Sediment Distribution: West-Central Florida Inner Shelf." 200:307-24.

Hartman, T. 2006. *Bivalve Shells of Florida*. Tampa: Anadara Press.

Katz, C. 1995. *The Nature of Florida's Beaches*. St. Petersburg, FL: Great Outdoors Publishing.

Puterbaugh, P., and A. Bisbort. 2001. *Florida Beaches*. 2nd ed. Berkeley, CA: Moon Publishing.

Witherington, B., and D. Witherington. 2007. *Florida Living Beaches: A Guide for the Serious Beachcomber*. Sarasota, FL: Pineapple Press.

7

Beaches of Alabama

THE coast of Alabama is not very long, but its beaches are almost all well developed. It extends from a portion of Perdido Key on the east through Dauphin Island across the mouth of Mobile Bay (figure 7.1). Like most of the northern Gulf Coast, the Alabama beaches have been severely eroded by tropical storms and hurricanes. Two recent hurricanes have resulted in major erosion of the beaches and destruction of built property: Ivan in 2004 and Katrina in 2005. These took place on a coast that was experiencing tremendous growth and development for the tourist industry. Obviously, good beaches are an integral part of this development, and these storms caused considerable loss of beach sand and tourism dollars.

Hurricanes are very destructive to beaches as well as the built environment. Nourishment is the primary way that beaches can recover from these storms. On the Alabama coast, nourishment took place along both Gulf Shores and Orange Beach in 2001 to mitigate the erosion of Hurricane Danny in 1997. The erosion from Hurricane Katrina required considerable nourishment to bring the beaches back for tourism. In 2006, what is one of the largest nourishment projects on the Gulf Coast was constructed with more than 7 million cubic meters of sand distributed along about 22 km of beach at a cost of $28 million.

This portion of the Alabama coast is a combination of developed sections and parks. There are two communities with multiple high-rise towers, Orange Beach and Gulfport. Excellent beaches extend throughout this area with generally small dunes on their landward side.

The Alabama portion of Perdido Key received beach nourishment as part of the large-scale project in 2001, as well as after Katrina. The beach here is now in very good shape (figure 7.2). This location is one of the best examples

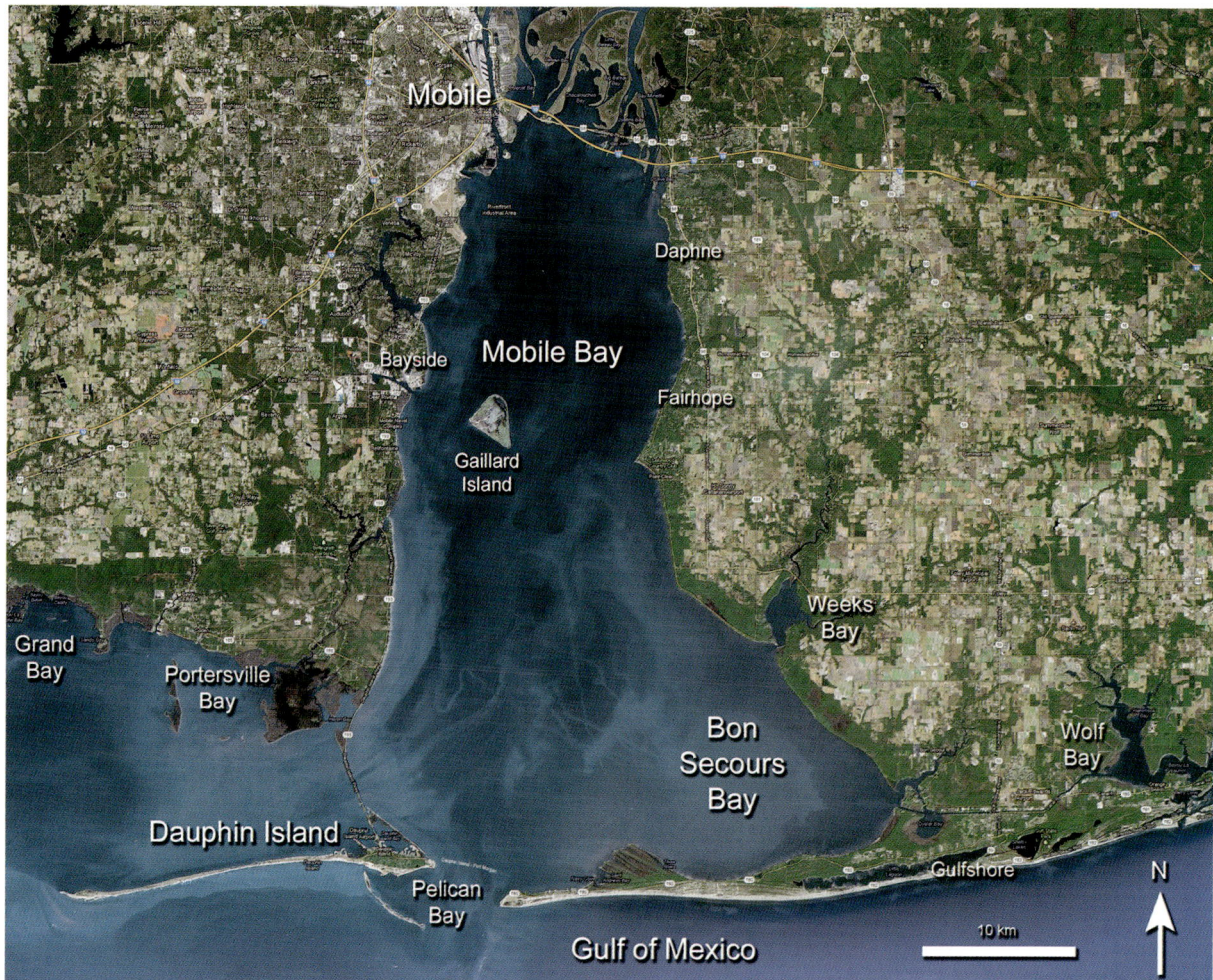

Figure 7.1. Satellite image of the Alabama coast showing the barrier islands. Photo courtesy NASA.

of how effective snow fences can be in helping to develop dunes. They are now buried with small dunes forming, and vegetation is becoming established. The large buildings in the distance front beaches that are equally well developed (figure 7.3). The heavily constructed community of Gulfport has excellent beaches that invite tourists for recreation (figure 7.4). This region had abundant oil on the beaches after the Deepwater Horizon spill in 2010. After considerable cleanup work, the beaches were returned to their essentially pure state in about a year.

Within the developed portion of the Alabama coast, there are parks with excellent beaches. The development of dunes is being encouraged through

Figure 7.2. View to the east along the beach at Gulf Island State Park on Perdido Key. In the distance is the Florida section of Perdido Key, which is part of the Gulf Islands National Seashore.

Figure 7.3. Beach on the state park looking toward the developed portion of Orange Beach. This part of the beach shows effective use of fencing to encourage dune growth.

fencing and vegetation, providing terrific coastal environments for both recreation and shoreline protection of the upland developed areas (figure 7.5). The western portion of this barrier system is only modestly developed, primarily by residential construction (figure 7.6). There are also extensive areas where the beach and adjacent environments are in pristine condition and receive few visitors (figure 7.7).

146 BEACHES ALONG THE GULF OF MEXICO COAST

Figure 7.4. Excellent highly developed beach in the tourist area of Orange Beach: (a) as it appeared before nourishment and (b) after nourishment.

Dauphin Island is the only Alabama barrier island west of Mobile Bay and is a mostly low, developed barrier that has experienced considerable modification by hurricanes. The western side of the Mobile Bay entrance is a combination of residential and undeveloped areas, whereas the eastern side is dense residential development. Dauphin Island has a nineteenth-century fort, Fort Gaines, at its eastern end where major shoreline changes have taken place (figure 7.8a). Erosion has caused the shoreline to move landward over the past few

Figure 7.5. Fencing along the backbeach placed to capture wind-blown sand to entice dune development.

decades, resulting in boulders as armor near the fort and tree stumps in the active beach near the eastern end of the island (figures 7.8b and 7.8c). Strong hurricanes have caused much erosion on the eastern end, which has left large groins isolated in the nearshore (figure 7.9).

Most of the sand in this system is concentrated on the eastern part of the island where dunes reach 15 m high, and considerable sand is moving onshore at the present time. Substantial accretion has been taking place on the eastern part of the island, just west of the erosion problems (figure 7.10). Because of the lack of mature dunes on the western two-thirds of the island, it is washed over by intense storms, resulting in major damage to both the natural and built

Figure 7.6. (a) Excellent beach in the western portion of the main Alabama coast province where residential development dominates. (b) Oblique aerial view of the natural barrier spit near the western end of the main Alabama barrier system. Photo b, courtesy R. Kirby.

Figure 7.7. Wide beach in area of residential development on the western part of this Alabama barrier island.

Figure 7.8. (a) Fort Gaines at the eastern end of Dauphin Island, (b) boulders at the shoreline to the west, and (c) tree stumps on the beach all show places where erosion is a problem.

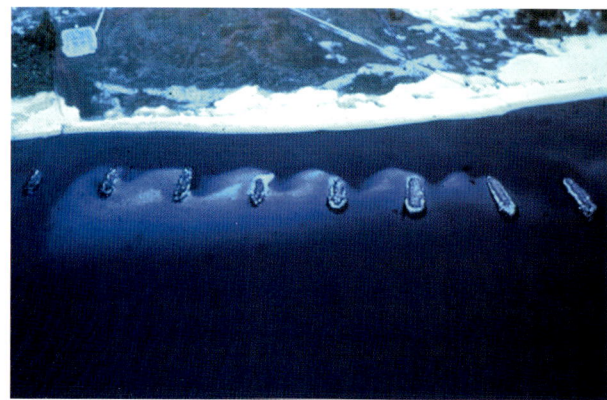

▲ Figure 7.9. Large groins that have been abandoned as the shoreline eroded. Photo courtesy D. Nummedal.

▲▶ Figure 7.10. Tremendous amount of accretionary sand in front of dune near the eastern end of Dauphin Island with more sand in the shallow nearshore zone in the distance.

▶ Figure 7.11. Dauphin Island, Alabama, after Hurricane Frederick in 1979. Photo courtesy D. Nummedal.

Figure 7.12. Washovers and overall island destruction as the result of Hurricane Katrina in 2005. Photo courtesy US Geological Survey.

island (figure 7.11). Hurricane Frederick in 1979 destroyed numerous residences, and the road was breached in several places. After major recovery of the beach and reconstruction of the houses along with the addition of many more, Hurricane Katrina struck the island in 2005. This storm essentially wiped out the houses and washed over the beach throughout much of the island (figure 7.12). A few years later the island had been completely rebuilt, and there are few indications of the previous devastation.

As time passes, the residential development on this beautiful island is progressing toward the west. The beaches here are wide and healthy but with some erosion problems locally (figure 7.13). There is more room for this type of development farther to the west where dunes are developing landward of the really nice beaches (figure 7.14). Unfortunately, Dauphin Island will always be vulnerable to hurricanes and tropical storms.

Figure 7.13. Erosion problems with armored protection at the western end of development on Dauphin Island.

Figure 7.14. Vast expanse of beach and coppice mounds toward the western end of Dauphin Island.

SUGGESTED READING

Canis, W. F., W. J. Neal, O. H. Pilkey Jr., and O. H. Pilkey Sr. 1985. *Living with the Alabama-Mississippi Shore.* Durham, NC: Duke University Press.

Otvos, E. G. 2004. *The Shores of Alabama and Mississippi.* Illustrated online chapters, in *The World's Coast.* Dordrecht, Netherlands: Kluwer Academic Publishers.

www.ngom.usgs.com/. A good place to access a wide range of US Geological Survey studies of the northern Gulf of Mexico.

8

Beaches of Mississippi

THE beaches of Mississippi are found on two distinctly different coasts: the mainland and four barrier islands that are several kilometers from the mainland (figure 8.1). None of the barrier islands is accessible by vehicle. For this reason and because they are mostly public land, the islands are pristine. A regular ferry schedule in spring and summer conveys people to West Ship Island, a federal park. The mainland beaches are among the most beautiful and best cared for along the entire Gulf of Mexico.

Beach nourishment has been common on the mainland of Mississippi. The first such major projects were after Hurricanes Ivan (2004) and Katrina (2005), which hit this coast very hard. A volume of 280,000 m³ of sediment

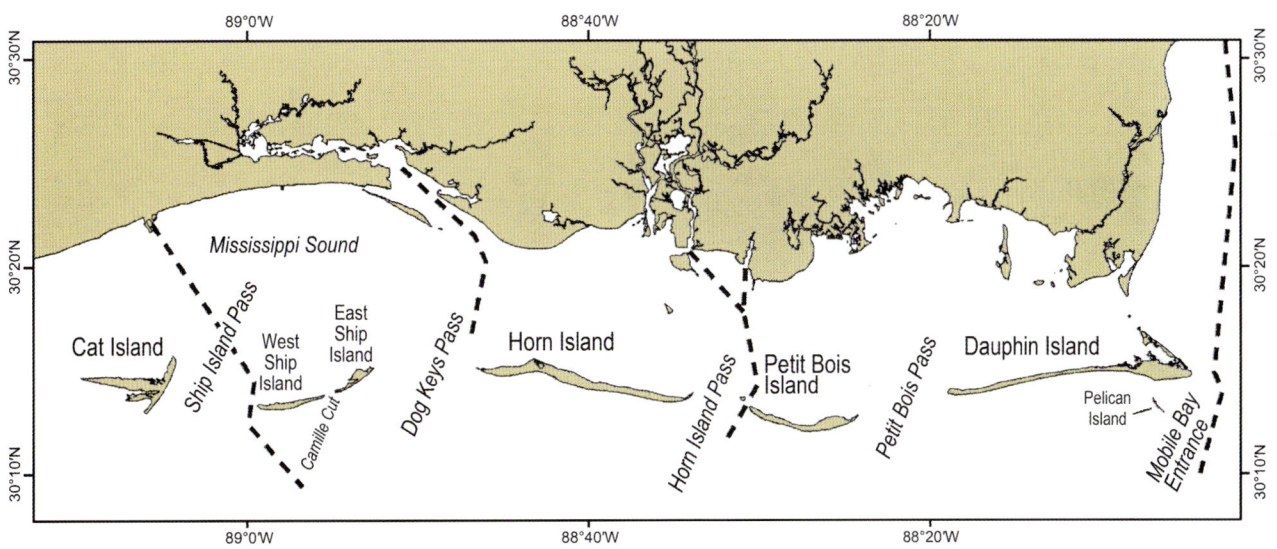

Figure 8.1. Map of Mississippi coast showing barrier islands far offshore and the mainland beaches. Dauphin Island, Alabama (on the right), is also included. Courtesy NASA.

was placed on the eroded beach at Pascagoula with the help of the federal government (figure 8.2). Farther to the west at Ocean Springs the beach is very well maintained. This area is dominated by fine sand with few shells. Wave energy is limited along this coast due to the offshore barrier islands that provide a level of protection. This low wave energy is evidenced by the vegetation near the strand line (figure 8.3). The Ocean Springs beaches also have groins to help maintain beach sediment.

Figure 8.2. Beach at Pascagoula nourished after the destruction of Hurricane Katrina in 2005.

Figure 8.3. Beach at Ocean Springs showing vegetation near the strand line and groins in the distance.

Mainland

The largest and most popular beaches along the Mississippi mainland coast are in the Biloxi Gulfport area, where tourism dominates because of several casinos. These wide, well-maintained beaches appear to be stable (figure 8.4). This condition extends for several kilometers along the coast. Because these nice beaches are in a major tourist area, the powers that be in the management system think that it is necessary and a good practice to regularly manicure the beach by raking it (figure 8.5). However, this practice is not good. Some beach sand is lost on a daily basis, and the beach profile is not allowed to equilibrate to a natural shape.

Another factor that can cause problems of a different type for the beach is the presence of storm drains throughout most of the beaches in the Biloxi-Gulfport area (figure 8.6). These drains can carry a wide variety of pollutants to the shoreline and into the shallow Gulf. As we continue toward the end of this reach, at Pass Christian the beach narrows, the shoreline becomes a bit irregular, and small structures are present (figure 8.7). There is also a vertical, poured concrete seawall at the landward side of the beach. This beach resembles that of Ocean Springs at the other end of this shoreline area.

One of the prettiest sections of beach on the Gulf Coast is located to the west at the small community of Waveland. This coastal community consisted of beautiful homes prior to Hurricane Katrina. The storm made landfall right at Waveland and essentially wiped out the entire community (figure 8.8). Virtually every home, mostly Victorian, was completely destroyed. This is a wealthy community, and the residents went to work to rebuild both the homes and the waterfront. They have built a seawall complex that is quite unique and attractive. It is also nicely planted and well maintained. The beach was nourished, and fencelike structures were constructed to enhance dune development (figure 8.9). This project was funded entirely locally through implementation of a "seawall tax." At the present time the houses are being rebuilt in the style of the original ones.

Barrier Islands

The barrier islands offshore of the Mississippi mainland are the farthest from land of any in the Gulf Coast, about 15 km. The four islands extend for more than 100 km along the Mississippi coast. Obviously, none of the barriers are in any way connected to the mainland. Their distance from land has also negated any development other than a couple of parks and a nineteenth-century fort.

Figure 8.4. Wide beach in the Biloxi-Gulfport area.

Figure 8.5. A wide beach in Biloxi where regular raking of the surface takes place, even in the nontourist season. This photo was taken in February.

The islands consist almost entirely of fine sand. The topography has limited relief, with small dunes in some areas. Unlike most coastal barriers, these have beaches on both the Gulf side and the landward side because of the fetch. Current vegetation is essentially all grass and small bushes; most trees were denuded and killed by Hurricane Katrina.

Figure 8.6. Beach in the Gulfport area with the large storm drain that empties into the Gulf.

Figure 8.7. The beach just east of Pass Christian (bridge in background) where the shoreline is irregular and vegetation is on the beach due to low wave energy.

The easternmost island in Mississippi is Petit Bois, a long and narrow barrier. It is curved with the convex side toward the Gulf. Sand Island is a small island on the west, separated from the main island by nearly a kilometer. The highest elevation on the island is only a meter or two above high tide. There is little difference between the beaches on either side of the island (figures 8.10 and 8.11).

▶ Figure 8.8. Virtually complete destruction by Hurricane Katrina of the residential community on the beach at Waveland. Photo courtesy NOAA.

▼ Figure 8.9. Nourished beach at the community of Waveland showing (a) the seawall complex and (b) the fence structure for dune enhancement.

◄ Figure 8.10. Low but wide beach on the bay side of Petit Bois Island. Each of the orange and white sections on the survey rod is 30 cm in this and subsequent photos. This photo and all of the rest in this chapter, unless otherwise noted, are courtesy Gulf Coast Spatial Center, University of Southern Mississippi.

◄ Figure 8.11. The beach on the open Gulf is wide and low with white sand.

Figure 8.12. Destruction of trees on Horn Island caused by Hurricane Katrina.

Horn Island is the longest of the Mississippi barriers with about 18 km of continuous, beautiful sand beaches on both sides. There are a pier and a ranger station on this island as part of the Gulf Islands National Seashore Park. Horn Island has more elevation than other islands and had numerous trees prior to Katrina (figure 8.12). The beaches show a distinct difference on the Gulf and bay sides (figures 8.13 and 8.14).

The next island, Ship Island, is actually now two islands, East Ship and West Ship, as a consequence of multiple hurricanes. Until Katrina, natural processes had repaired this separation. Now it is almost 5 km wide. Katrina also caused significant damage to historic Fort Massachusetts on the northern side of West Ship. At the present time there is a huge federal government–funded project to repair the cut between the islands. When Hurricane Katrina hit East Ship Island, the beaches were overwashed and sand transported toward the mainland. The trees were defoliated and killed (figure 8.15). The landward

BEACHES OF MISSISSIPPI 161

▲ Figure 8.13. Gulfside beaches on Horn Island are subjected to substantially more wave energy than on the landward side.

◀ Figure 8.14. The low-energy beaches show irregular shorelines.

▲ Figure 8.15. Beach on the Gulf side of East Ship Island showing destruction of trees by Hurricane Katrina.

▶ Figure 8.16. Low-energy landside beach on East Ship with low-tide flats.

Figure 8.17. Oblique aerial of Fort Massachusetts on West Ship Island. Photo courtesy National Park Service.

Figure 8.18. Natural beach on the Gulf side of West Ship with numerous widespread coppice mounds showing evidence of accretion.

side of the island is rather low energy with essentially tidal-flat-like surfaces (figure 8.16).

West Ship is the only one of the Mississippi Sound barriers that has abundant visitation. The presence of Fort Massachusetts, coupled with the availability of a regular ferry service, brings many thousands of tourists to the island annually (figure 8.17). The Gulf beach here is very wide and covered with small coppice mounds (figure 8.18), indicating that sediment is being added

Figure 8.19. A shallow trench in a West Ship Island beach showing dark, heavy mineral layers. Photo courtesy Ping Wang.

Figure 8.20. Beach area on West Ship used by visitors for recreation.

to the beach and dunes will be forming in the absence of hurricanes. Heavy minerals are commonly displayed on the beaches of West Ship Island, and thin layers may be incorporated in the beach stratigraphy (figure 8.19). The visitors to West Ship will find amenities that are absent on all of the other Mississippi barriers (figure 8.20).

Figure 8.21. Eroding beach and upland areas on Cat Island; most of the erosion is the result of Hurricane Katrina.

Figure 8.22. Severe damage caused by Hurricane Katrina, which destroyed the trees as well as much of the shoreline.

Figure 8.23. Good sand beach on the Gulf side of Cat Island.

The most complex and interesting of the Mississippi barriers is Cat Island, which is actually a complex with two different origins. The "T shape" of the island owes its origin in part to a lobe of the Mississippi delta and in part to wave processes of late Holocene shoreface conditions. The essentially north-south part of the island originated from reworking of a delta lobe. The beaches on the eastern side of the island are eroding badly due to storm activity and the absence of new sediment (figure 8.21). Hurricane Katrina had a major effect on the island, killing trees and eroding upland areas (figure 8.22). There are good beaches here, however (figure 8.23).

SUGGESTED READING

Canis, W. F., W. J. Neal, O. H. Pilkey Jr., and O. H. Pilkey Sr. 1985. *Living with the Alabama-Mississippi Shore*. Durham, NC: Duke University Press.

Otvos, E. G., 2004. *The Shores of Alabama and Mississippi*. Illustrated online chapters, in *The World's Coast*. Dordrecht, Netherlands: Kluwer Academic Publishers.

www.barrierislandsms.com. Excellent website about many aspects of the Mississippi barrier islands.

www.mybearhawk.com/flying/islands.html. "Island Hopping" website that consists largely of aerial photos of the Mississippi barrier islands.

www.ngom.usgs.com/. A good place to access a wide range of US Geological Survey studies of the northern Gulf of Mexico.

9

Beaches of Louisiana

THE beaches of Louisiana are probably the least attractive and least visited on the entire Gulf Coast. This is primarily the result of the Mississippi River and delta dominating the coast of this state, and the fact that only one barrier island, Grand Isle, is accessible by vehicle. There are several kilometers of mainland beaches that are fairly popular.

The river system produces a huge volume of fine sediment that dominates the coast. Much of this fine sediment remains suspended as it leaves its distributary channels, producing muddy water that is not attractive to tourists but is very important for the community of organisms that lives along this coast. The bulk of the sediment discharged by this fluvial system either remains in the Louisiana coastal zone or is transported westward. Some moves offshore and is deposited on the Mississippi Fan in deep water.

The beaches in Louisiana are limited to narrow, low barrier islands formed by reworking of abandoned lobes of the delta and to the low-lying chenier plain of the western portion of the state. In both settings the sand that composes the barriers and their contained beaches is perched on thick mud. As a result, the sand is sinking. The rapid rate of sea-level rise along this coast is causing a problem for the long-term existence of the barriers, and the size and elevation of the barriers make them very vulnerable. In fact, they are eroding rapidly and are regularly washed over by storms that spread the sand landward. Several attempts have been made to halt or reduce this erosion. Unfortunately, the future of these barriers, and therefore the beaches, is not very promising.

Beaches on Barriers of Delta Lobes

There are several abandoned and relict lobes of the Mississippi delta in addition to the one that is present now (figure 9.1). As sea level rose during the postglacial transgression of the Gulf shoreline, there was a reworking of the abandoned sediment lobes of the river by waves, especially during storms. This reworking concentrated the sand at the outer margin of the lobes and carried the fine sediment away by tidal and wave-generated currents. These concentrations of sand have become barrier islands with natural beaches.

The Chandeleur Islands are an excellent example of "lobe barriers" (figure 9.2). They are located on the St. Bernard lobe east of the modern delta lobe. It formed from 4000 to 2000 years before present. These islands are low but have vegetation, and they are washed over fairly regularly. They also were breached at numerous locations (figure 9.3). These barriers have never been occupied and are rarely visited by tourists. Because of the rapid erosion combined with the rapid rate of sea-level rise, it is possible that the Chandeleurs will not last through this century. A long berm has been constructed in front of these islands to protect them.

The only barrier island accessible from land is Grand Isle (figure 9.4), and it is the only one that has any residential development (figure 9.5). This barrier

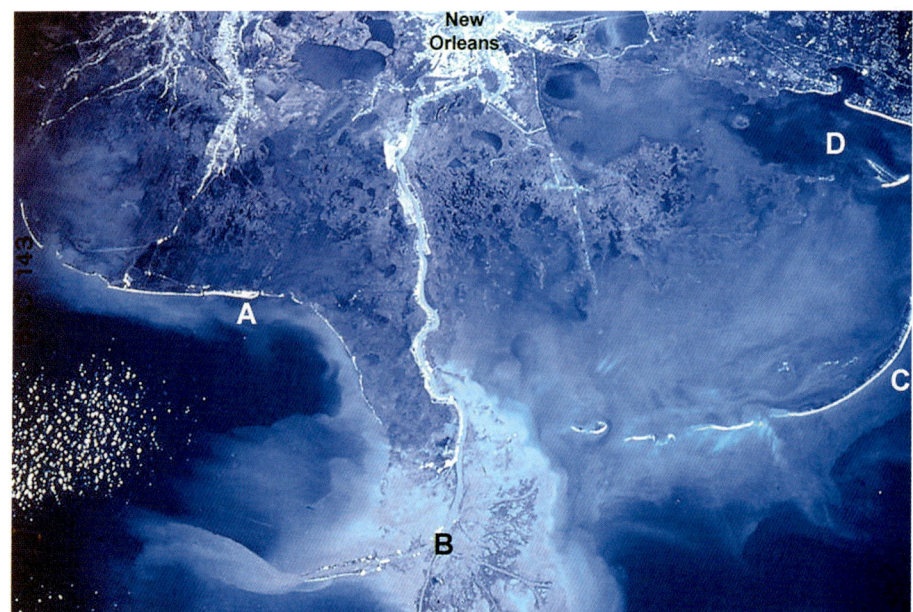

Figure 9.1. Aerial image of the Mississippi delta complex showing multiple barrier islands that have formed from the reworking of relict delta lobes: A and C are barriers reworked from deltaic lobes, B is the modern active lobe, and D is mainland beaches. Photo courtesy NASA, modified by J. Aber, Emporia State University.

BEACHES OF LOUISIANA   169

▲ *Figure 9.2. Oblique view of the Chandeleur Islands prior to Hurricane Katrina, which caused considerable erosion.*

▲ *Figure 9.3. Part of the Chandeleur Islands where storms regularly breach the barriers. The erosion here and the channels are the result of Hurricane Katrina. Photo courtesy US Geological Survey, 2007*

island developed on the outer portion of the LaFourche lobe, which was active from 2500 to 800 years before present. Like all the Louisiana barrier islands, this one is also in jeopardy. It is rapidly being eroded and washed over during intense storms. It is, however, the most likely of the barriers to persist because of some well-designed offshore breakwaters. This is very important because Grand Isle is critical to the petroleum industry as a base for offshore activities and a residence for hundreds of offshore workers. Nearby Port Fourchon is a major support facility for the petroleum industry. Grand Isle is also a fairly important tourist destination because it is the only accessible beach in central Louisiana.

Beginning at the eastern end of Grand Isle, at the state park, there is abundant evidence of efforts to protect and manage the beach. As a consequence of its vulnerability and its importance to the state of Louisiana, it is being protected from erosion. A terminal groin has been constructed to keep beach sediment from moving into the channel to the east. Fences have been placed to trap sand for dunes, and grasses planted to trap and hold sand in the dunes (figure 9.6).

At the present time on the eastern half of the island there are multiple segmented, offshore breakwaters designed and located to reduce erosion on the barrier (figure 9.7). It is apparent that they are helping because there are well-developed beaches landward of them and sediment has accumulated in the form of salients along the shoreline (figure 9.8). These salients show that the breakwaters are reducing wave energy.

Figure 9.4. Infrared image showing Grand Isle, a developed but very fragile barrier island. Courtesy Coastal Protection and Restoration Authority of Louisiana.

Figure 9.5. Oblique aerial view of Grand Isle, Louisiana, showing the development. Photo courtesy US Geological Survey.

Figure 9.6. Looking to the eastern end of Grand Isle at the state park, showing a terminal groin, sand fences, and plantings in the dunes.

BEACHES OF LOUISIANA   171

Figure 9.7. Oblique view of the beach at Port Fourchon, west of Grand Isle, showing multiple segmented breakwaters for beach protection and preservation. Photo courtesy US Geological Survey.

Figure 9.8. Large salient that has formed landward of one of the segmented breakwaters at the eastern end of Grand Isle.

▼ *Figure 9.9. (a) Dune and beach system near the eastern end of Grand Isle where breakwaters are present and (b) the same subject in the western portion of the island where breakwaters are absent.*

The breakwaters extend about halfway along the island. A comparison of beach/dune systems that are protected by the breakwaters and those that are not shows no discernible difference (figure 9.9). The engineers who designed the system seem to have had it figured out! It is interesting to note that the material that concentrates in the swash zone under low-energy conditions because of the breakwaters looks at first glance like oil because of the location, but in fact it is plant debris from peaty material in the back-barrier marsh.

Farther to the west of Grand Isle is Isles Dernieres, one of the barriers that developed on a relict delta lobe but is not accessible to vehicle traffic. This island is very vulnerable to destruction. The combination of sediment compaction, sea-level rise, and severe storms is rapidly causing the demise of this barrier island system (figure 9.10). It has gone from one large and one small island to four very small islands in the span of about 150 years.

Beaches Formed on the Cheniers

The western part of the Louisiana coast is dominated by a chenier plain (figure 9.11). The term *chenier* refers to the low ridges that generally have relatively tall vegetation. These features are the result of reworking of dominantly muddy sediment by wave action, leaving shelly sand ridges. As time passes, these ridges accumulate and the shoreline progrades into the Gulf. This process has been going on since sea level reached near its present position. The result of these processes has produced a sandy shoreline that contains beaches. Although this coastal reach is vulnerable to hurricanes and sea-level rise, it is attractive to people who are looking for a place to fish and relax at the beach. These attributes have produced the small community of Holly Beach (figure 9.12).

There were problems with beach erosion before the destructive hurricanes of the early twenty-first century. In the early 1990s a series of segmented breakwaters was installed to mitigate erosion by reducing wave energy at the shore (figure 9.13). These structures did not solve the problem, so the beach was nourished in 2002. Nourishment sand came from about 30 km offshore. A total of about 2.5 million cubic meters of borrow material was placed on the beach and in the form of dunes over about 12 km of coast at a cost of $45 million. The combination of offshore breakwaters and beach nourishment produced an excellent beach with tombolos (salients) extending from the mainland to the structures (figure 9.14). This condition was operating quite well until Hurricane Rita hit in 2005.

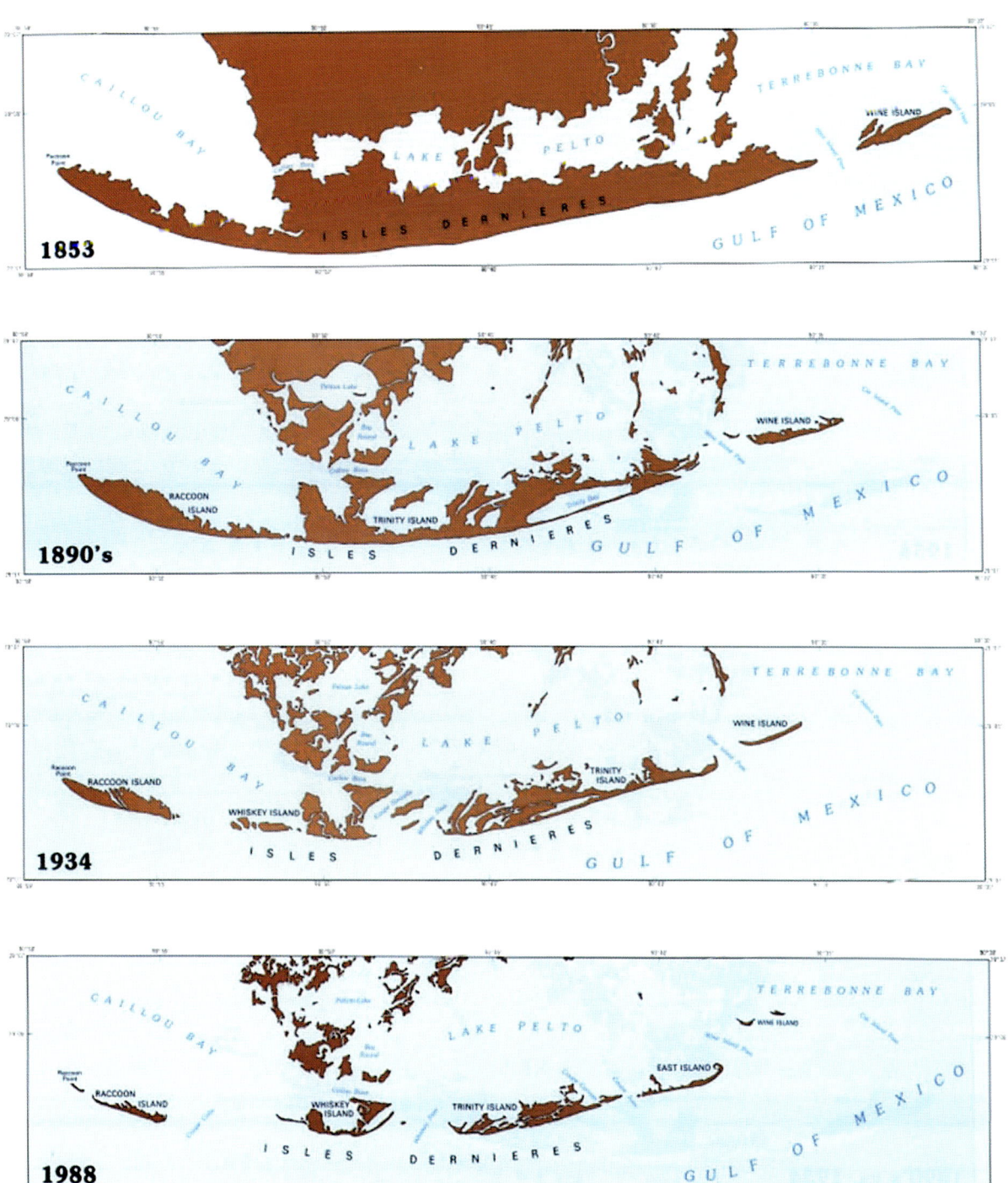

Figure 9.10. Sequence of maps of Isles Dernieres over 150 years showing the decrease in surface area with the future in major jeopardy. From S. Penland, C. Zganjar, K. A. Westphal, P. Connor, A. Beall, J. List, and S. J. Williams, Shoreline Posters of the Louisiana Barrier Islands, 1885 to 1996, *US Geological Survey, Open File Report 03-398, 1996.*

Figure 9.11. (a) Map and (b) aerial view of the chenier plain area of western Louisiana where mainland beaches are common. Photo a, from J. Byrne, D. O. Le Roy, and C. M. Riley, "The Chenier Plain and Its Stratigraphy, Southwestern Louisiana," Transactions of the Gulf Coast Association of Geological Societies 9 (1959): 237–60.

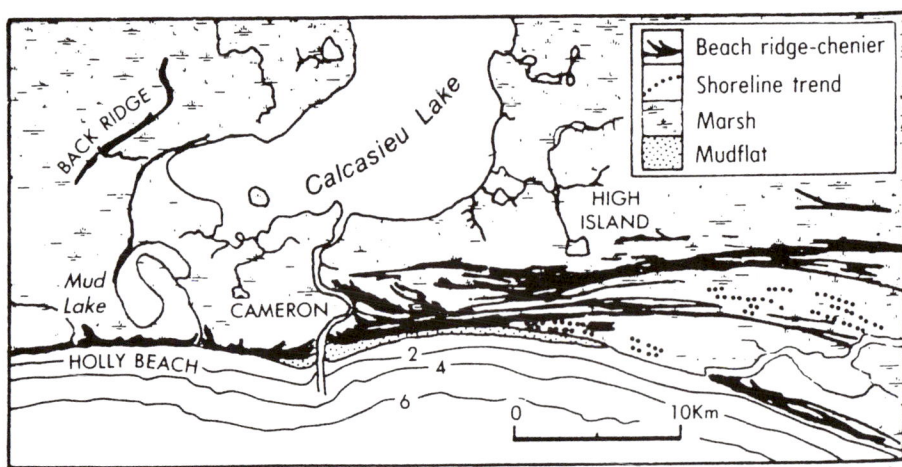

This community has been subjected to severe hurricanes that have essentially wiped it out. Rita was especially devastating (figure 9.15). Even though these tragedies destroyed the community, it has been rebuilt and the beach system is now (2012) in good shape. Fences have been installed to trap sand and build dunes on the backbeach (figure 9.16). There are small dunes already forming (figure 9.17), and the community has been rebuilt with hopes of sustaining its existence (figure 9.18). The beach is in very good shape, and the road is being protected from erosion.

Farther west, away from the development at Holly Beach, there are serious erosion problems. The highway is even in jeopardy. As a consequence,

BEACHES OF LOUISIANA 175

Figure 9.12. Oblique aerial view of the community of Holly Beach before Hurricane Rita in 2005. Photo courtesy US Army Corps of Engineers.

Figure 9.13. Offshore breakwaters that were installed in the early 1990s to protect Holly Beach. From M. Mouledous, Holly Beach Sand Management Summary Data and Graphics, Louisiana Department of Natural Resources CS-31 (Baton Rouge: LDNR, 2003). Photo by Victor Monsour.

Figure 9.14. Beach nourishment and breakwaters that protect the Holly Beach community. From M. Mouledous, Holly Beach Sand Management Summary Data and Graphics, Louisiana Department of Natural Resources CS-31 (Baton Rouge: LDNR, 2003). Photo by Victor Monsour.

Figure 9.15. Post–Hurricane Rita (2005) showing the complete destruction of the entire community of Holly Beach. Courtesy US Geological Survey.

Figure 9.16. Fence placed on the backbeach to trap sand and build a dune system.

segmented breakwaters have been placed in shallow water offshore (figure 9.19), and fences have been put on the backbeach in hopes of trapping sand (figure 9.20), but the beach is so narrow that there is no dry sand to blow back to begin dune formation. This Long Beach area may be good for fishing but not for water recreation.

Figure 9.17. Small dunes that have been constructed and are covered with vegetation for stability.

Figure 9.18. View of Holly Beach community across the beach as it appeared in December 2011. All houses are new and constructed under current zoning regulations.

Figure 9.19. Narrow beach near Long Beach where offshore breakwaters have been installed and fences have been placed to help maintain the beach.

Figure 9.20. Very narrow, steep beach showing considerable erosion and a vulnerable adjacent highway.

SUGGESTED READING

Ritchie, W., K. A. Westphal, R. A. McBride, and S. Penland. 1992. *Coastal Sand Dunes of Louisiana: An Inventory (Plaquemines, Isle Dernieres, Chandeleurs)*. Baton Rouge: Louisiana Geological Survey.

Williams, S. J., S. Penland, and A. H. Sallinger, eds. 1992. *Atlas of Shoreline Changes in Louisiana from 1853-1989*. U.S. Geological Survey, Misc. Invest. Series I-2150A. Reston, VA: USGS.

www.ngom.usgs.com/. A good place to access a wide range of US Geological Survey studies of the northern Gulf of Mexico.

10

Beaches of Texas

THE Texas coast is essentially a continuum of beaches with tidal inlets scattered throughout (figure 10.1). With few exceptions, these beaches are on barrier islands that are no more than 7000 years old. Mainland beaches are present between Follets Island and Matagorda Peninsula. This is a distinctly wave-dominated coast with low to moderate energy and a mean annual wave height of about 0.5 m. Because the prevailing wind is from the southeast, much of the coast experiences a northeast-to-southwest longshore transport of sediment. In contrast, from the Rio Grande mouth north to an area known as "Big Shell" in central Padre Island, the longshore transport is in the opposite direction. The passage of cold fronts between October and March produces strong wind from the north that blows generally offshore and dissipates wave energy for a few days each year. This process and the relatively strong prevailing wind have resulted in this coast being considered wind dominated, a more specific category of wave-dominated coasts.

Bolivar Peninsula

The easternmost portion of the Texas coast is part of the chenier plain complex of western Louisiana and East Texas. There are wide beaches just west of the Sabine River that front an extensive low-elevation coastal plain (figure 10.2). Although there is abundant sediment as evidenced by the wide beach, there has been no development of dunes along this part of the coast. Much of this sediment originally comes from Mississippi River discharge.

Farther to the west the beach becomes narrow and then absent, and the coast is erosional overall. In fact, over the past couple of decades the highway

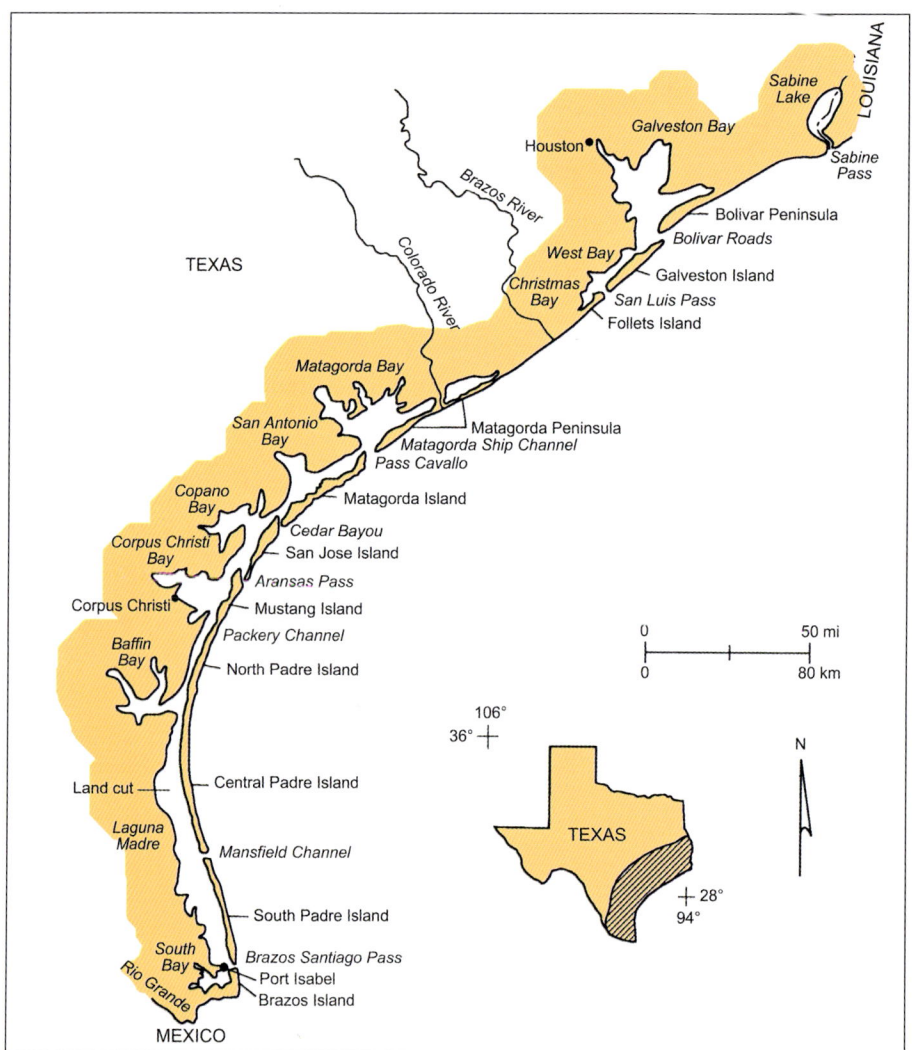

Figure 10.1. Outline map showing the barrier islands of the Texas coast. After R. A. Morton, "Texas Barriers," in Geology of Holocene Barrier Island Systems, ed. R. A. Davis, pp. 75–114 (Heidelberg, Germany: Springer-Verlag, 1994).

along this area has been removed by shoreline retreat. There are still several kilometers of highway from the community of High Island to the west, with a moderately wide beach gulfward of the road. It is apparent that this area is also subjected to severe weather conditions, as evidenced by the low concrete wall located between the active beach and the road (figure 10.3). The wall is placed here to prevent beach material from being transported on to the highway during severe storms. This beach has a high concentration of shells, both modern ones, and oyster shell and rock that was reworked from older sediments in the shoreface offshore.

▲ Figure 10.2. (a) Wide, low beach at Rim State Park just west of the mouth of the Sabine River. (b) Low, duneless area landward of the beach.

◀ Figure 10.3. Low concrete wall placed to prevent beach material from washing onto the highway on the Bolivar Peninsula. The beach here has a high shell content.

As one continues along the Bolivar Peninsula to the community of Crystal Beach, there is additional evidence of erosion problems. Fencing originally designed for capturing snow in higher latitudes has been placed at the landward side of the beach with the intent of capturing sand and stabilizing dunes (figure 10.4). The western end of the peninsula actually appears more like a barrier island. The beach becomes quite wide, and small dunes have developed (figure 10.5). The entrance to Galveston Bay interrupts the shoreline here. The entrance includes the ship channels leading to both the Ports of Galveston and Houston, where much of the tonnage that crosses the Gulf of Mexico comes and goes.

Galveston / Folletts Island

The largest and most extensive development on the Texas coast is on Galveston Island. The city of Galveston was the largest in Texas in the nineteenth century and was a major port. One of the most spectacular events of the early days

Figure 10.4. Fences on the landward side of the beach are placed to try to capture sand and nourish the small foredunes.

Figure 10.5. Western end of Bolivar Peninsula showing (a) wide beach looking toward entrance to Galveston Bay and ship channel, (b) wide beach toward the east, and (c) dunes that are about 1.5–2.0 m high on landward side of the beach.

Figure 10.6. The Corps of Engineers raised the level of the gulfward portion of the town of Galveston to accommodate the seawall and avoid flooding in the future. Photo courtesy US Army Corps of Engineers.

in Texas was the hurricane of 1900, which claimed 8,000 lives and resulted in major changes to help protect the city from similar events in the future. One was elevating the city by about 2 m by dredging and filling under buildings that were left after the storm (figure 10.6). This increase in elevation extends only a few blocks back from the shoreline, which was the city of that day. The other primary event was the construction of the Galveston seawall, an engineering marvel of its day. This seawall was originally just over 5 km long but has been extended to 16 km. It is 5.2 m high and 4.9 m thick at its base.

The easternmost portion of the island, adjacent to the ship channel, has shown accretion of more than a kilometer to the width of the beach in less than a century (figure 10.7). It is a bit hard to realize that the Galveston seawall now rests so far from the present shoreline and that waves were crashing at its base about 100 years ago (figure 10.8). This large fillet of sediment extends for a few kilometers to the west where the seawall becomes exposed to the present coastal processes. A fairly nice beach is present near the end of the fillet (figure 10.9). Large groins were added to keep the beach sediment from moving westward along the wall (figure 10.10).

One of the important design aspects of this wall was the curved face that, along with the large boulders at its base, helps dissipate the wave energy. Groins

▲▲ Figure 10.9. Beach in front of the seawall near the western end of the sediment fillet.

▲ Figure 10.10. Large groins attached to the seawall. These structures were placed after the construction of the original wall and are designed to keep sediment from moving along the shoreline.

▲▲ Figure 10.7. Looking gulfward from the Galveston seawall to the present shoreline, a distance of more than a kilometer.

▲ Figure 10.8. Boulders at the base of the seawall that were originally in the zone of wave attack. Now they are well behind the shoreline, and the present shoreline is a long way to the west.

of large granite boulders have also been placed at the wall to further reduce the incident wave energy (figure 10.11). As can be seen in the photo, there is no dry beach present even though beach nourishment had taken place here several years before the photo was taken. A large beach nourishment project was undertaken for the Galveston area in 1995 because there had been little or no beach in front of the seawall. Unfortunately, the project was not very successful, and the sand was eroded in a rather short time because the nourishment material was too fine grained.

At the western end of the seawall, there is a significant problem. This is typically referred to as the "end effect" in coastal structure terminology. At the end of a protective structure such as a seawall, there is commonly a significant amount of erosion and shoreline retreat. There is considerable protection at the end of the Galveston wall, but the shoreline is displaced almost 100 m (figure 10.12). When this wall was built, there was no development west of its end; now there is considerable building all the way to the end of the island.

◀ *Figure 10.11. Seawall as it looks along most of its extent along the highly developed portion of the city of Galveston.*

▶ *Figure 10.12. The end of the seawall at Galveston (a) shows a marked displacement with unprotected shoreline, (b) has large, well-placed granite blocks, but (c) shows much erosion from wave attack. Aerial photo is from 1971.*

Figure 10.13. (a) Beach with little dune development, which is typical of the western end of Galveston and Follets Islands. (b) Beach with small dunes on the landward side that have developed since Hurricane Ike in 2008.

The western end of Galveston Island and the adjacent Follets Island have similar beaches and landward portions (figure 10.13). They are, however, separated by San Luis Pass, a rather large tidal inlet. The beaches in this area are quite vulnerable and have been attacked by severe storms, most recently Hurricane Ike in 2008. Plastic geotubes filled with sand have been placed to help maintain the beach, but with little success. There is little dune development in this area (figure 10.14), and there are several washover channels cutting the island (figure 10.15); elevations overall are quite low. Crystal Beach at the western end of Follets Island has experienced considerable loss of homes from Hurricane Ike but has recovered quite well.

The fact that the beaches in this area are quite wide; have a flat, nearly horizontal surface; and are composed of very fine, well-sorted sand makes them perfect for vehicular traffic. They are, in fact, very popular driving pathways, and people take camping gear and use the beach as a public campground. It is permitted, but there are no facilities except the blue waste cans.

Surfside, near the eastern end of Follets Island, is a community that has been racked by storms, the latest of which was Hurricane Ike. Rebuilding has been extensive, and the beach is now in good shape except toward the eastern part of the development. One of the techniques for mitigating beach and dune problems is using old Christmas trees to act as sand traps and thereby enhance the growth and development of dunes (figure 10.16). Recent nourishment (2012) has been necessary to curtail further erosion and provide some recreational beach (figure 10.17).

Figure 10.14. Wide beach and small dunes that characterize eastern and central Follets Island. Note the car tracks showing the driveability of this type of beach.

Figure 10.15. Washover channel from Hurricane Ike (2008); these are very common on western Galveston Island and Follets Island.

Sargent Beach Area

There is an absence of barrier islands for about 65 km from the town of Freeport across the mouth of the Brazos River to the Matagorda Peninsula. The Brazos River originally emptied into the Gulf at what is now the port at Freeport. It was rerouted in 1929 to provide a stable port without problems of flooding in order to accommodate the huge petroleum industry in the Freeport area. The new mouth of the Brazos has not developed a true delta although there has been substantial sediment delivered to this area. It has now been reworked into chenierlike, welded ridges (figure 10.18). The area is often characterized as a wave-dominated delta.

The area at Sargent Beach has been developed as a weekend community since the 1950s and has been a site of severe beach erosion since that time. In fact, two blocks of residences have been lost in about a half century. As early as the 1970s the shoreline was characterized by exposed marsh deposits (figure 10.19). The Intracoastal Waterway is located less than 200 m from the Sargent Beach shoreline, and there was concern about uncontrolled erosion causing impact on this important waterway.

▲ Figure 10.16. Christmas trees placed on the landward edge of the beach to trap sediment and thereby begin construction of sand dunes.

▲▶ Figure 10.17. Recent beach nourishment at Surfside to thwart further erosion and provide some recreational beach area.

This situation continued until the Corps of Engineers constructed a low-lying protective structure to hold the shoreline in place. The structure was a combination of a low-lying concrete wall and a complex of specially placed granite blocks (figure 10.20). This structure has not only done the job of holding the shoreline in place but is nearly completely covered by sand so that it does not present the negative aesthetic appearance that is the case with many similar structures.

Matagorda Island / San José Island / Mustang Island

Matagorda Peninsula and Island extend for about 75 km and are interrupted near the middle by the Colorado River (figure 10.21). This coastal reach is nearly all natural except for a small residential development just east of the mouth of the river. The beach here is wide and composed of sand with scattered shells. Small coppice mounds are present in front of a low foredune complex (figure 10.22). Two of these barrier islands provide rather major contrasts in some respects and a lot of similarity in others. Matagorda is a nature preserve and is pristine, whereas San José (St. Joseph) is a private island with virtually no development except for a cattle ranch. Mustang Island is relatively developed and home to Port Aransas, a rapidly growing small city that is popular with both fishermen and tourists.

San José is 34 km long and separated from Matagorda Island at the northern edge of Cedar Bayou, a large, complex washover feature (figure 10.23). There is a main channel at this location that closes naturally from longshore transport of sand. This channel is important to the health of Aransas Bay on the landward side of the islands and is scheduled to be dredged in 2013.

▲ Figure 10.18. Mouth of the Brazos River where it was relocated in 1929. It does not have a distinct delta, but sand is being accreted on the western or downdrift side of the river mouth.

◄ Figure 10.19. Sargent Beach in the early 1970s, showing considerable erosion in this land view, evidenced by the relief and irregular nature of the shoreline.

San José is owned by the Bass family of Fort Worth and is an operating cattle ranch. The ranch headquarters, an airstrip, and a small harbor represent the only development on an otherwise pristine barrier island (figure 10.24). The island has excellent beaches and dunes, and its adjacent waters are popular fishing locations both on the Gulf and the bay sides (figure 10.25). Beaches on San José Island are not only pristine but also quite beautiful (figure 10.26). The only public access to the island is via the jetty boat, a commercial vessel that runs like a taxi from Port Aransas on adjacent Mustang Island.

Figure 10.20. Close-up of (a) granite blocks shortly after completion of the structure in 1998 and (b) their current buried condition, demonstrating that the design was done well.

Port Aransas is on the northern end of Mustang Island at Aransas Pass, a huge, structured tidal inlet (figure 10.27). This inlet is the entrance to the Port of Corpus Christi several kilometers into and across Corpus Christi Bay. The community of Port Aransas is growing rapidly and is a prime tourist destination on this part of the Gulf Coast (figure 10.28). This northern end of

BEACHES OF TEXAS 191

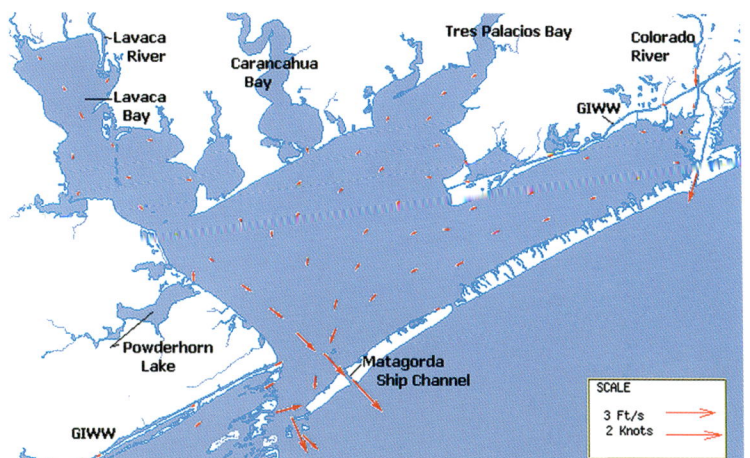

Figure 10.21. Map showing the Colorado River delta as it progrades through Matagorda Bay and empties into the Gulf as it cuts through the Matagorda Peninsula. GIWW is the Gulf Intracoastal Waterway.

Figure 10.22. The beach on the Matagorda Peninsula is well developed with (a) a wide, dry beach and (b) a developing dune system. The dark material in the backbeach is Sargassum that washed up during the late winter of 2012. Photos courtesy Gay Hejtmancik.

Figure 10.23. Oblique aerial view of Cedar Bayou, which separates Matagorda Island from San Jose Island. The black line marks the position of the main channel that is the boundary between the two islands.

Mustang Island must be accessed by a state-run, free ferry that is part of the Texas Department of Transportation. The community houses the University of Texas Institute of Marine Science, a research and teaching campus affiliated with the main campus in Austin.

Fishing and the beaches are the major attractions for tourists on this island. Fishing is done anywhere along the shore, including from the jetties, and in the bays, offshore and the surf. Beach access is typically by vehicle with the "beach road" being maintained by the local highway personnel. An annual parking permit is available throughout the town. Generally the beach is easily driveable with a conventional two-wheel drive vehicle, but storms may cause changes that prohibit traffic. One should be aware of access and road conditions.

Mustang Island beaches are excellent for recreation and may become crowded, especially on holidays and summer weekends. Amenities such as trash barrels and Porta Potties are available, and overnight parking is permitted for campers. Surfing is popular under proper conditions, but those are infrequent throughout the year. The best place is near Horace Caldwell fishing pier, but there may be issues between fishermen and surfers.

The beach on the northern end of the island is modified to accommodate large numbers of tourists with designated parking lots. The beach has been

BEACHES OF TEXAS 193

Figure 10.24. Beach on San José Island showing what is almost a complete tree, the type of debris that washes onto beaches in this area after storms.

Figure 10.25. Oblique aerial view of the pristine beach and dune complex on San José Island.

▶ *Figure 10.26. Beach and eroded dune complex on the southern part of San José Island.*

▼*Figure 10.27. Aerial view looking gulfward along Aransas Pass with San José Island on the left (north) and Mustang Island on the right (south). The University of Texas Marine Science Institute is in the lower right corner. Note the difference between the sand fillets on each side of the inlet.*

graded landward to provide a wide beach and parking space, so the dunes are now landward of where they should be in a natural condition (figure 10.29). Beyond this area the beach is natural, more narrow but still quite wide and a maintained vehicular thoroughfare. The beach and dunes are natural and in excellent condition (figure 10.30). Along this long stretch of Mustang Island are spots where development is taking place (figure 10.31)

The beach on the southern portion of Mustang Island is being severely graded to facilitate traffic. This practice is not a good one because it creates an unnatural beach profile that can accelerate erosion (figure 10.32). The southern boundary of Mustang Island is rather artificial in that it is the recently constructed Packery Channel. This is the site of a former natural inlet that had been closed for many years. Now it is a very popular place for fishing and recreating.

Figure 10.28. Oblique aerial view across Mustang Island showing the southern portion of Port Aransas with the Corpus Christi ship channel in the background (black line).

Figure 10.29. Developed beach where widening has taken place and parking lots are designated. This is not wise beach management, but it serves the purpose for tourist development.

Figure 10.30. Natural beach and dune on Mustang Island that contrasts with the image in figure 10.29.

▲ Figure 10.31. Oblique aerial view showing where new development is encroaching on Mustang Island in order to take advantage of the excellent beach.

◄ Figure 10.32. Beach near the southern end of Mustang Island where considerable "road" grading has taken place to help with beach traffic. Such practices are not good beach management; the beach profile should be left to develop naturally.

North Padre Island

The northern end of Padre Island is developed heavily for residential and tourist accommodations. It is a very popular area because of its easy access and good beaches. Immediately south of Packery Channel is a large seawall protecting high-rise condominiums and a hotel (figure 10.33). This beach requires nourishment, which is supplied by dredging sand from Packery Channel. This is the only nourished beach along this coast for more than 150 km. Surfing is also popular, especially around Bob Hall Pier, a favorite fishing location. These beaches are accessible and usually very good for recreation and fishing, but there may be times when the brown algae *Sargassum* covers the beach, making it inhospitable for tourists (figure 10.34).

A few kilometers down the island is the entrance to the Padre Island National Seashore, one of the first and largest of the national seashore parks. The main visitor center is at Malaquite Beach. There are camping sites nearby and other amenities. Driving on the beach is easy for the first 8 km or so, and then a sign warns that only four-wheel drive vehicles should proceed beyond that location. This part of Padre Island is very popular with surf fishermen and visitors who want to get to remote parts of the barrier. On the way down the island

Figure 10.33. Wide nourished beach just south of Packery Channel that was built with sand from dredging the inlet. Erosion has been a problem in part because of the large seawall in this area.

Figure 10.34. Common situation along this part of the Texas coast when large volumes of Sargassum *wash onto the beach, which occurs primarily in the spring.*

one notices the large dunes. There are old and buried tidal inlets along the way. Yarborough Pass was open until 1950 and is now the adjacent beach area to an excellent surf-fishing spot (figure 10.35).

The rest of Padre Island is rather similar, down to the jetties at Mansfield Pass (figure 10.36), except for the areas called "Little Shell" and "Big Shell." The natural profile includes wide beaches, coppice mounds, and foredunes (figure 10.37). Because of the pristine conditions and the absence of people, it is a common place for campers who like to be alone (figure 10.38). The two shell-dominated beach sections each extend for several kilometers beginning about 15 km after the four-wheel drive sign. Both of the sections are essentially all-shell beaches. It is easy to recognize them from your moving vehicle because the surface becomes soft and sometimes is difficult to pass through, even with four-wheel drive (figure 10.39). As the names imply, one is covered with small shells and pieces, and the other has larger shells (figure 10.40). The jetties at Port Mansfield extend more than 100 m into the Gulf. An aerial view of this inlet shows that considerably more beach sand is being transported toward the north than to the south (figure 10.41).

Figure 10.35. Oblique aerial view of the Yarborough Pass area where an open channel flowed in the mid-twentieth century.

▼ *Figure 10.36. More typical condition of Padre Island beaches, which are wide, soft sand surfaces. Photo courtesy J. W. Tunnell.*

Figure 10.37. Common appearance of beach profiles in this area are a wide beach, coppice mounds, and foredunes.

Figure 10.38. This part of Padre Island is a good place for campers who want some isolation. Photo courtesy US National Park Service.

Figure 10.39. It is apparent when the shell beaches are reached by the soft sediment and deep vehicle tracks. Photo courtesy Alistair Lord.

Figure 10.40. (a) Beach and surf at Big Shell and (b) close-up of the shell gravel sediment.

Figure 10.41. The jetties at Mansfield Pass show northward accumulation on the southern side. Photo courtesy US Department of Agriculture.

South Padre Island

Most people consider only the southernmost, highly developed part of Padre Island as "South Padre," but in this discussion all of this barrier island south of the Mansfield jetties will be included under that name. South Padre is about 60 km long, the vast majority of which is natural and beautiful barrier island with a healthy beach. Because most of this portion of Padre Island is accessible only by four-wheel-drive vehicle, it is quite remote. The rare visitor beyond the paved road is generally a surf fisherman.

Beginning at the jetties at Mansfield Pass the most dramatic feature of the shoreline is the large offset on each side of the jetties. South-to-north longshore transport of sand has caused a huge accumulation on the south jetty (figure 10.42). The beach to the south is sorted sand with scattered shells and occasional *Sargassum* (figure 10.43).

In general, the beach along all of the pristine portion of South Padre Island is wide with sand, scattered shells, and seasonally, *Sargassum* (figure 10.44). There are commonly wheel tracks from the fishermen's vehicles unless a storm has caused the swash to wash them out. Sand is abundant, and dunes are generally well developed. The wide, dry beach and onshore winds give rise to abundant coppice mounds that eventually become dunes (figure 10.45). In the event of severe storms, the dunes will be reached by waves and erosion will occur. The result is a scarp formed on the Gulf side of the dunes. These scarps may persist for many years before they become healed (figure 10.46).

The community of South Padre Island is at the southern end of the barrier

Figure 10.42. Wide beach from the jetty at Mansfield Pass to the south.

Figure 10.43. Typical foreshore beach on South Padre Island with fine sand, scattered shells, and Sargassum.

Figure 10.44. Common beach profile on South Padre Island with abundant Sargassum on a wide beach.

Figure 10.45. Excellent coppice mounds on this part of Padre Island indicate a large sand supply that is blowing to the backbeach.

Figure 10.46. Dunes that were scarped by erosion from Tropical Storm Dolly in 2008 and four years later are being healed

island and has become a very popular coastal resort destination for college students for spring break and families for both summer and winter vacations (figure 10.47). With this dense development, some of which has not been well planned, there have been problems with beach erosion. Fortunately, the combined efforts of the community, the State of Texas, and the federal government have addressed these rather effectively. The obvious best answer is to nourish the beach.

The first beach nourishment was conducted in 1988 by placing dredged sediment from the main ship channel, Brazos Santiago Pass, which needs to be dredged on a regular basis. From 1988 to 1995 the dredge spoil was placed in a berm offshore of the town of South Padre Island at a depth of 8 m as a feeder system to the eroding beach. That was a slow and inexpensive method of nourishment. Beginning in 1997 nourishment took place using the more conventional approach of placing the sand directly on the beach (figure 10.48). The dredged sediment has been pumped along the beach to the construction site, about 8.3 km to the north. The most recent of these projects was constructed in 2011 at a cost of more than $4.5 million. Beach nourishment is an ongoing need for this community. Now there is a wide beach for tourists that is well maintained and well used (figures 10.49 and 10.50).

Figure 10.47. Oblique aerial view of the city of South Padre Island, a major coastal resort area. Photo courtesy Jackie Reeves-Wilson.

Figure 10.48. Nourishment in progress at South Padre Island in 2011. The large pipe has carried the dredged sediment for several kilometers along the beach. Photo courtesy Jacques Halle, World Wide Press.

▲ Figure 10.49. Steep beach along the developed portion of southern Padre Island.

▶ Figure 10.50. Very nice beach that is quite popular with the tourists at South Padre Island.

SUGGESTED READING

Anderson, J. B. 2007. *Upper Texas Coast*. College Station: Texas A&M University Press.

Hayes, M. O. 1965. Sedimentation in a Semiarid, Wave-Dominated Coast (South Texas) with Emphasis on Hurricane Effects. PhD diss., University of Texas, Austin.

Morton, R. A. 1994. "Texas Barriers." In *Geology of Holocene Barrier Island Systems*, ed. R. A. Davis. Heidelberg, Germany: Springer-Verlag.

Morton, R. A., and J. H. McGowen. 1980. *Modern Depositional Environments of the Texas Coast*. Guidebook GB 20. Austin: University of Texas, Bureau of Economic Geology.

Morton, R. A., and R. L. Peterson. 2006. *South Texas Coastal Classification, Maps—Mansfield Channel to the Rio Grande*. US Geological Survey, Open File Report, 2006-1133. Reston, VA: USGS.

Morton, R. A., and W. A. White. 1980. *A Guide to the Geology, Natural Environments and History of the Texas Barrier Island*. Guidebook GB 17. Austin: University of Texas, Bureau of Economic Geology.

11

Beaches of Mexico and Cuba

OVERALL, the Gulf Coast of Mexico is relatively unpopulated and therefore rather pristine. Areas around population centers of Veracruz and Tampico are exceptions. This chapter considers some of the major places where people will visit. The discussion of the Mexican coast of the Gulf terminates in the Cancún vicinity.

The Cuban shoreline is not well known and is frequented only by non-US citizens at this time. The northern coast just east of Havana is the most popular place to visit and has excellent beaches. There are two styles to the shoreline zone in Cuba, and each is discussed.

Mexico

The beaches are much the same in northern Mexico as they are in South Texas. The back-barrier lagoon here is also called Laguna Madre. Overall, the beaches of Mexico are fine sand and are terrigenous except for the area of Campeche Bay and the Yucatán Peninsula, where carbonate skeletal debris dominates beach sediment. This material is coarser than that on the terrigenous beaches. In the area between the two distinct sediment types the beaches are dominated by a mixture of quartz and carbonate debris, giving the sediment a bimodal texture.

Much of the information on Mexican beaches comes from the book *Environmental Atlas of the Gulf of Mexico*, in the chapters by Patricia Moreno-Casasola and Arturo Carranza-Edwards et al. (see suggested reading at end of the chapter).

Terrigenous Beaches

The beaches in the State of Tamaulipas have their origin primarily in the deltaic sediments deposited at the mouth of the Rio Grande. This barrier island system is quite similar to that on the US side at South Padre Island except for the level of development. A narrow barrier island is separated from the mainland by Laguna Madre (figure 11.1). The beaches tend to be very wide, up to 250 m just south of the federal border. They slope gently toward the Gulf and are a combination of quartz and feldspathic sand with scattered shells and shell debris. Dune development is limited, and the beach is accretional. The width of the dry beach decreases toward the south, and the slope steepens until at Miramar, where the beach is less than 10 m wide with a steep foreshore.

The beaches in the area of La Pesca north of Miramar have abundant shell material and are bimodal in texture because of it (figures 11.2 and 11.3). These steep beaches have eroding clay dunes on their landward side in La Pesca (figure 11.4). The clay in these dunes comes from fluvial sediment carried to the Gulf by the Soto Marina River.

The State of Veracruz has a wide range of coastal types. In the northern part of the state the beaches are wide and sand dominated, similar to those in Tamaulipas (figure 11.5). Farther south the beaches have a distinctly different composition due to the Quaternary volcanic activity. Rocky sections dominate

Figure 11.1. Oblique aerial view of the northern Tamaulipas coast with a narrow, pristine barrier fronting Laguna Madre. Photos in figures 11.1–11.16 courtesy J. W. Tunnell.

Figure 11.2. Beach surface showing a wide, gently sloping profile with a few scattered large pieces of limestone.

Figure 11.3. Close-up of a typical shell concentration common on the northern Tamaulipas coast.

Figure 11.4. Steep, narrow beach that extends to erosive clay dunes, La Pesca area, Tamaulipas.

Figure 11.5. Fairly wide, sandy beach near the border of the States of Veracruz and Tabasco at Sanchez-Magallanes.

Figure 11.6. Bedrock shoreline reflecting the nature of the volcanic coast at Veracruz

this coastal region (figure 11.6). As a result, both volcanic rock fragments and glass are common in the beach sediment. Boulder beaches are common and are generally somewhat narrow and steep with widths less than 50 m (figure 11.7). The beach sediment consists of boulders, some nearly a meter in diameter. This material is very coarse but tends to be well sorted (figure 11.8). These are unlike any other beaches on the Gulf.

Figure 11.7. Some beaches in the State of Veracruz are narrow and steep with very coarse sediment.

Figure 11.8. Close-up of moderately sorted cobbles and boulders from a Veracruz beach. The scale bar is 39 cm.

In the State of Tabasco the beaches become more narrow with the maximum being less than 30 m. The sediments are the most terrigenous dominated of this coast, partly because of the abundance of fluvial discharge and the inhibition of carbonate production due to turbid water. The sand here is also fine to very fine grained, the finest on the Mexican coast. Such beach sand has some potential for economic recovery of heavy minerals such as ilmenite, which is used in paint, and magnetite, an ingredient used in the coating industry for large boilers and other pieces of equipment.

An interesting factor in the Tabasco beaches is that turtles, which nest throughout the Mexican beaches, do not nest on this part of the coast, probably because of the very fine and, in some places, muddy texture of the beach sediment. Turtles apparently cannot or do not like to dig in this material.

Carbonate Beaches

Carbonate beach sediment becomes common in the State of Campeche. This sediment is essentially all shell material. These beaches are coarse sand and gravel in texture and are accepted by turtles for nests. This area is far from any fluvial input and is on the margin of the Yucatán Peninsula, a carbonate platform. The beach composition becomes about 50% carbonate in the area of Isla del Carmen (figure 11.9). The Campeche beaches are generally narrower than those northward in Tabasco.

As one moves up the Campeche coast, the beach is gone and the shoreline is Tertiary limestone (figure 11.10). This limestone is fairly fine grained and extends well into the upland part of the coast. The sand beaches on the Campeche coast are all carbonate shell debris. There are a few indications of erosion, including a small scarp at the landward end of the beach and some

Figure 11.9. Carbonate shell sand beach on Isla del Carmen in the State of Campeche.

Figure 11.10. Shoreline dominated by Tertiary limestone in an absence of any sand on the Campeche coast.

Figure 11.11. Small scarp on the landward end of the carbonate sand beach on the Campeche coast, indicating that erosion can take place on this low-energy shoreline.

very primitive groins to disperse wave energy and mitigate erosion (figures 11.11 and 11.12). Along one section of the coast mangroves grow at the open-water shoreline, much like what is present on the southwestern coast of Florida (figure 11.13).

Figure 11.12. Another indicator of shoreline erosion shown by the somewhat steep, narrow shoreline with primitive groins to disperse wave energy and mitigate erosion.

The State of Yucatán has good beaches, but most are narrow and rather steep (figures 11.14 and 11.15). All the beaches are carbonate shell material. This is probably the most energetic part of the Yucatán Peninsula. The swash area here may have a concentration of shells that is quite spectacular (figure 11.16).

Around the Yucatán coast to the northeastern corner is the State of Quintana Roo and the end of the Mexican coast of the Gulf of Mexico. Here is the major tourist destination of Cancún, a place where few Mexican citizens recreate but millions of foreign tourists do. Vast numbers of high-rise hotels and condominiums face small, well-manicured beaches (figure 11.17). The water is beautiful, but the beaches may be quite narrow and steep (figure 11.18). Hurricanes have impacted this coast over the past few decades and have resulted in various types of protective structures being built (figure 11.19). Unlike most of the Gulf of Mexico, the sand removed from the beach by storms does not return to Cancún beaches during calm weather conditions. Strong tidal currents in the channel between the mainland and nearby Isla Mujeres tend to carry the eroded beach sand to the south. A recent large-scale beach nourishment project has provided wide beaches.

Figure 11.13. A portion of the Campeche shoreline has mangroves on the open coast: *(a)* close-up of red mangrove (Rhizophora mangle) and *(b)* mangal shoreline.

Figure 11.14. Beach at San Bruno is steep and narrow, typical of the northern Yucatán Peninsula.

Figure 11.15. A modestly developed Yucatán beach with a narrow but beautiful beach.

Figure 11.16. Shell concentration such as can be found on the foreshore portion of beaches near San Bruno, Yucatán. Scale bar is 10 cm.

Figure 11.17. Beach manicuring as performed on a regular basis on Cancún beaches by these raking machines. Photo courtesy D. Kinley and S. Kinley.

Figure 11.18. Typical, rather narrow, steep beach along the developed Cancún shore. Photo courtesy D. Kinley and S. Kinley.

Figure 11.19. Protection for the beaches is common in Cancún because of major losses of beach sand due to the impact of hurricanes. Photo courtesy D. Kinley and S. Kinley.

Cuba

The part of Cuba that is included within the Gulf of Mexico is essentially the northwestern tip of the coast to the end of the Varadero Beach area, about 150 km east of Havana. There are three distinct morphologic characters to this region: (1) the western high-relief portion that is sparsely occupied and little known, (2) the zone just west of Havana that is a combination of carbonate sand and limestone reef rock, and (3) the sandy beaches of Varadero Beach. The latter is the most extensive and is the main tourist region of Cuba. This area is a major part of the economy of this country.

The westernmost portion of the coast is high relief and dominated by bedrock shorelines with some small pocket beaches. This coast is rather remote except for a few small communities that surround a few small estuaries. This coastal geomorphology extends to the Havana area.

West of Havana is an area dominated by rocky shorelines and sand beaches where locals and other Cubans spend recreational time at the beach (figure

Figure 11.20. Typical sand beach in the area west of Havana. Low dunes are common landward of the active beach.

Figure 11.21. Beach in this area with small limestone outcrop (arrow) exposed in the swash zone.

11.20). This region is only about 20 km long and is a mixture of carbonate sand and reef rock. The sand comes from the breakup of carbonate animal skeletal material offshore and from the waves eroding the old reefal material at the shoreline. There are rock exposures scattered throughout the sand beaches (figure 11.21). These old reefs are from the Pleistocene, the time of the last highstand of sea level about 120,000 years ago when sea level was 3–5 m above its present

Figure 11.22. Late Pleistocene reef that now forms the shoreline. The coral reef platform was about 3–5 m below sea level about 100,000–120,000 years ago: (a) the platform and (b) the eroded shoreline. Photo b, courtesy A. C. Hine.

level. Reefs were abundant in this area, and the rocky portion of the shoreline was alive and about 2–3 m below sea level at that time (figure 11.22).

The beaches east of Havana are relatively steep and not very wide. There are small dunes on their landward margin with sparse vegetation. It is apparent that coastal management is not well organized here because there are many pathways through the dunes where vegetation is absent. This condition allows wind erosion and keeps the dunes from becoming continuous and well developed.

Eastward is the Varadero Beach region, the best and biggest coastal resort area in Cuba (figure 11.23). This is essentially the end of the Gulf Coast of Cuba. Varadero Beach is a straight section of coast that extends to the northeast and diverges from the mainland of Cuba. This narrow coastal province is anchored by late Pleistocene carbonates, both eolianites (lithified dunes) and reef rock. Sediment is abundant, and the beaches are both well developed and well maintained with many hotels, some large and fancy and others small and basic.

Figure 11.23. Sunset on the western portion of Varadero Beach where most hotels are small and older than on the eastern end, where most foreign tourists visit. Photo courtesy A. C. Hine.

SUGGESTED READING

Brogdon, D. R. 1954. "Beach Sands of the Gulf Coast—Tamaulipas, Mexico." Master's thesis, University of Texas, Austin.

Carranza-Edwards, C., L. Rosales-Hoz, M. C. Chavez, and E. M. de la Garza. 2004. "Environmental Geology of the Coastal Zone." In *Environmental Analysis of the Gulf of Mexico*, ed. K. Withers and M. Nipper, pp. 351–72. Corpus Christi: Texas A&M University–Corpus Christi, Harte Research Institute of the Gulf of Mexico. Translated from Spanish and available on the HRI website, www.harteresearchinstitute.org.

Casasola, P. M. 2004. "Beaches and Dunes of the Gulf of Mexico: An Analysis of the Current Situation." In *Environmental Analysis of the Gulf of Mexico*, ed. K. Withers and M. Nipper, pp. 302–13. Translated from Spanish and available on the HRI website, www.harteresearchinstitute.org.

Glossary

accessory minerals. Mineral species that are typically present is small quantities in a sediment sample.

antidunes. A bedform that results from upper flow regime conditions that is low amplitude with the crest oriented toward the upstream direction.

backshore (backbeach). The nearly horizontal landward part of the beach that is typically dry.

beach. The zone from the low-tide line to the landward distinct change in topography or composition.

beach cusps. Small, somewhat triangular-shaped shoreline features that can be coarse sediment or sand like the adjacent beach.

beach nourishment. Addition and sculpting sand to reconstruct the beach after erosion.

beachrock. Layered rock lithified in the beach environment and composed of very shelly sediment, usually limited to low latitudes.

bedforms. The regular perturbation of the surface produced by currents, e.g., ripples and aqueous dunes.

bimodal sediment. Sediment that comprises two populations of distinctly different grain sizes.

boulder. A sediment grain that is >256 mm (about 10 inches) in diameter.

chenier. A small ridge generally composed of shells and shell debris that is reworked by storms from muddy coastal deposits.

clay. A type of silicate mineral; a grain size (1/256 mm).

cleavage. Plane of weakness in the crystallography of a mineral.

cobble. A sediment particle that is 64–256 mm in diameter.

cold front. A weather system in which cold high pressure and strong northerly wind follows low pressure.

coppice mounds. Small accumulations of wind-blown sand that lie just seaward of the foredunes. They are essentially incipient dunes.

current crescent. Scour around a pebble or shell on the beach that produces a direction-oriented structure.

diurnal tides. Tidal cycles with one high tide and one low tide each tidal day.
ebb-tidal delta. The sediment accumulation at the seaward end of a tidal inlet.
eustasy. Global change in sea level.
fetch. The distance over which wind blows on a water basin.
flood-tidal delta. The sediment accumulation at the landward end of a tidal inlet.
foreshore. The seaward portion of the beach on which there is uprush and backwash of water by the last wave motion.
frontal system. A moving weather system in which a cold air mass is adjacent to a warm air mass.
gravel. Refers to all grain sizes larger than sand (>2.0 mm).
heavy minerals. Minerals that have a specific gravity of greater than 2.85 gm/cm^3 and that are a minor constituent in most sediments.
horn. That part of a beach cusp that protrudes seaward.
hurricane. A cyclonic low-pressure storm with minimum wind velocity of 75 mph.
jetty. A shore-normal structure that is constructed to stabilize a tidal inlet or channel.
longshore current. A coastal current that is produced by the refraction of waves as they move through the nearshore environment.
longshore sandbars. Elongate accumulations of sand that parallel the shoreline and are only tens to hundreds of meters from it.
mixed tides. Tidal cycles that are a mixture of diurnal and semi-diurnal tides during a lunar month. The Gulf of Mexico experiences these tidal cycles.
mixed-energy barrier island. A barrier island that develops as a result of the combination of both wave-generated processes and tide-generated processes. The product is a so-called drumstick barrier.
mixed-energy tidal inlet. An inlet that results from a combination of both tidal and wave processes. It has a substantial ebb-tidal delta with a distinct terminal lobe and a well-developed flood-tidal delta that resembles a horseshoe crab in configuration.
morphology. The process-response systems that cause changes.
mud. Sediment that is the combination of silt and clay-sized particles.
neap tide. Tidal cycle that has minimal change in water level during a lunar month.
nearshore. The coastal environment that extends from the shoreline across the bar and trough topography.
ooids. Sand-sized particles that are carbonate rings surrounding a nuclei that can be either biogenic or terrigenous.

overwash. The process whereby waves carry sediment across the beach and onto the backbarrier or beyond.

pebble. Sediment particle size between 4 and 64mm.

perched beach. Term applied to small nourished beach terminated at each end by a groin, generally of longard tubing.

plunging breaker. A wave that breaks essentially instantaneously and is generally produced by a breaking swell wave.

progradation. Seaward movement of shoreline by the addition of sediment.

rhythmic topography. Exhibited by a shoreline that appears as a low-amplitude sine wave.

ridge and runnel. A low-relief topographic feature in the intertidal beach that is the result of high-energy conditions. This feature is ephemeral, lasting a few weeks to a few months.

rip current. A narrow current that flows seaward across the surf zone through a saddle in the longshore bar and is dangerous for swimmers.

ripples. Small-scale bedforms that can be formed by waves or currents.

roundness. A grain-shape parameter that describes the smoothness of the surface.

saddle. Low area in the crest of a longshore sandbar.

salient. An accumulation along the shoreline caused by a protective feature that reduces waves and longshore currents.

sand. Grain size of sediment between 0.625 and 2.0 mm.

sand shadows. Sand accumulations in the lee of an object, such as a shell or pebble/cobble, that show wind direction.

sea wave. A wave, generally rather steep with a peaked crest, that is currently under the direct influence of wind.

sediment budget. The spatial and temporal distribution of sediment as it moves in and through various coastal environments.

sedimentary structures. Features that have repetitions or special appearance due to sediment movement or organism movement.

semi-diurnal. Tidal cycle with two high and two low tides each tidal day.

setup. The very small increase in water level at the shoreline caused by onshore waves under normal conditions.

shoreline rhythm. The low amplitude, sine-curve shape of the shoreline.

silt. Sediment grain size between 4 microns and 0.0625 mm (62.5 microns).

sorting. The distribution of grain sizes within a sediment sample. Narrow distribution is well sorted, and a wide range of particle sizes is poorly sorted.

sphericity. The three-dimensional shape of a sediment particle. All axes the same length are spherical, whereas axes of a range in lengths are not.

spilling breaker. A breaker, generally produced by breaking sea waves, that breaks over some time and distance, like liquid spilling out of a container.
spring tide. Tidal cycle that has maximum change in water level during a lunar month.
storm surge (storm tide). The increase in water level produced by the friction of strong wind over the water surface. It may extend several meters along the coast.
surf zone. The zone of breaking waves in the nearshore environment.
surging breaker. The most shoreward breaking waves when there is a landward surge of the wave as it breaks.
swash marks. Very thin, arcuate lines of sediment on the foreshore marking the most landward encroachment of a given wave. They are concave seaward.
swash zone. The portion of the beach foreshore that is traversed by uprush and backwash of waves.
swell wave. A deep-water wave that is not directly under the influence of wind. It generally has low wave height and long wavelength.
terminal groin. A shore-normal structure at the end of a nourishment project to keep sediment from being transported out of the designated area.
terrigenous. Derived from land; typically referring to sediment.
tidal prism. The volume of water that passes through an inlet during a tidal cycle.
tide-dominated inlet. A tidal inlet characterized by a large ebb-tidal delta with long channel-margin linear bars and an absence of a terminal lobe.
tropical depression. Cyclonic low-pressure system with a maximum wind speed of 39 mph.
tropical storm. Cyclonic low-pressure system with a wind speed between 39 and 74 mph.
undertow. Return flow along the bottom from the shoreline seaward.
washover apron. Coalescing of multiple adjacent washover fans.
washover fan. A fan-shaped accumulation of sediment produced by sediment-laden currents that transport material across and to the back of a barrier island.
wave-dominated barrier island. Long, narrow barriers with abundant washover fans formed primarily by wave-generated processes.
wave-dominated inlet. A tidal inlet that has a very small ebb-tidal delta and an unstable channel.

wave period. The amount of time in seconds it takes for one wavelength to pass a point.

wave reflection. The result of a wave impacting a vertical obstruction such as a seawall or a steep foreshore of the beach.

wave refraction. The bending of waves as they move through the shallow nearshore zone to the shoreline.

Index

accessory minerals, 54, 55
    heavy minerals, 55, 56, 164, 213
Alabama coast, 143, 144, 148
Anclote Key, Florida, 38
Anna Maria Island, Florida, 120, 122
antidunes, 60, 61
aragonite, 55
attached bar, 23

back-barrier, 128
backbarrier bay, 31, 117
backbarrier marsh, 172
backbeach, 27, 28, 31, 61, 62, 95, 117, 119, 122, 147, 174, 205
Bahamas, 7
barrier island, 36, 154, 155, 180
Bay of Fundy, 49
beach cusps, 29, 30
beach management, 124, 124, 196, 197
beach manicuring, 219
beach morning glory, 98, 100
beach nourishment, 81, 82, 85, 153
beach profile, 32, 155, 194, 201
beach sediment, 113, 117
beach textures, 47
beachrock, 33, 34, 120
bedforms, 59
bedrock, 212, 220
Big Bend, Florida, 35, 105
Big Sarasota Pass, Florida, 45, 120
Big Shell, 199, 202
Biloxi, Mississippi, 155, 156
bimodal, 50, 51, 66, 67, 76, 209
bivalve, 91, 93, 94
Bolivar Peninsula, Texas, 179, 181, 182
borrow material, 82, 109
boulders, 149, 184, 212, 213
Brazos River, Texas, 187, 189
breaking waves, 10, 21
breakwater, 72, 126, 171, 172
Brownsville, Texas, 88
bucket dredge, 82

Bunces Pass, Florida, 43, 125
burrow, 94, 97, 99

Caladesi Island, Florida, 40, 128
calcium carbonate, 55, 58
Campeche, Mexico, 35, 214, 215
Cancun, Mexico, 216, 217, 219
Captiva Island, Florida, 112, 113, 114
carbon dating, 33
carbonate beaches, 214
Caribbean Sea, 7
Casey Key, Florida 119
Cat Island, Mississippi, 165, 166
Cayo Costa Island, Florida, 113, 115
Cedar Bayou, Texas, 192
Chandeleur Islands, Louisiana, 168, 169
chemical sediments, 58
chenier plain, 167, 174
Christmas trees, 186, 188
Christmas trees, 90
clay, 47, 54, 55
clay dunes, 209, 210, 211
Clearwater Beach Island, Florida, 128
Clearwater Pass, Florida, 127
Clearwater, Florida, 87
cleavage, 52
cobbles, 213
cold front, 4
Colorado River delta, 191
continental shelf, 7
coppice mounds, 28, 199, 201
Corps of Engineers, 65, 81, 183
Corpus Christi, Texas, 104
Crystal Beach, Texas, 181, 186
Cuba, 35
current crescents, 62
cusps, 62
cyclone 9

Dauphin Island, Alabama, 32, 146, 149, 150, 151

Deepwater Horizon spill, 144
delta lobe, 168
diurnal tide, 16
dragline and barge, 84
dredge, 82
    dredging, 198
    hopper dredge, 82, 84, 85
    suction dredge, 82, 84
drumstick barrier, 37, 40, 45, 128
Dry Tortugas, Florida, 105, 106
Dunedin Pass, Florida, 128
dunes, 88, 147, 169, 171, 182, 189, 193, 195, 196, 203
    foredunes, 89, 199

ebb shield, 42
ebb-tidal currents, 42
ebb-tidal delta, 37, 42, 43, 45, 82, 125
Egmont Key, Florida, 122, 124
erosion, 19, 56, 66, 118, 126, 152, 186, 206, 214, 215, 216

feldspar, 54, 55, 58
fence, 176, 183
    sand fences, 169, 170
    snow fence, 89
fetch, 9
    fetch limited, 9, 103
fillet, 76, 194
flood-tidal delta, 37, 42
Florida coast, 7, 25, 28, 76, 95, 97, 109, 110, 113
Florida Keys, 104
Florida panhandle, 4, 7, 39, 52, 103, 179, 181
Follets Island, Texas, 186, 187
foraminifera, 55, 56, 57
foredunes, 89, 199
foreshore, 26, 27, 31, 51, 85, 92, 218
Fort Massachusetts, Mississippi, 163
frontal system 5

231

Galveston Bay, Texas, 181
Galveston Island, Texas, 182, 185–187
Galveston, Texas, 70, 71, 104, 182, 185
Gasparilla Island, Florida 116
geotextile tubes, 68, 87
ghost crab, 97, 99
ghost shrimp, 94
glaciers, 18
grain shape, 51, 52
grain size scale, 48
Grand Isle, Louisiana, 167, 168, 170, 171
gravel, 47
groin, 74, 77, 123, 183
Gulfport, Alabama, 143
Gulfport, Mississippi, 155, 156

Havana, Cuba, 220, 223
heavy minerals, 55, 56, 164, 213
High Island, Texas, 180
Holly Beach, Louisiana, 172, 174–177
Holocene, 166
hopper dredge, 82, 84, 85
Horn Island, Mississippi, 161
Hudson Bay, 18, 19
hurricane, 7, 107
Hurricane Danny, 143
Hurricane Elena 1985, 25
Hurricane Frederick 1979, 147, 150
Hurricane Ike 2008, 8, 186
Hurricane Ivan 2004, 153
Hurricane Katrina 2005, 153, 156, 158, 160, 162, 165, 166
Hurricane Pass, Florida, 41
Hurricane Rita 2005, 174, 176
hurricanes, 174

Indian River, Delaware, 80
intertidal beach, 26
Intracostal Waterway, 187
Isla del Carmen, Mexico, 214
Isla Mujeres, Mexico, 216
Isles Dernieres, Louisiana, 172, 173

Jacksonville, Florida, 74
jetty, 76, 78, 79, 194

Key Largo Limestone, 105–107
Key West, Florida, 107, 108

Laguna Madre, Mexico, 209
Lake Michigan, 66, 68
Lido Key, Florida, 120
limestone, 210, 211

limestone, 210, 211, 214, 215, 221
Little Shell, 199
Long Beach, Louisiana, 178
Long Key, Florida, 125, 126
Longboat Key, Florida, 120–122
longshore bar, 12, 21, 22
longshore currents, 12, 37
longshore transport, 45, 72
Louisiana, 72, 73
Louisiana coast, 167, 172
lunar cycle, 14, 15
lunar tide, 16

Maine coast, 16
Manasota Key, Florida 113
mangroves, 109, 128, 215, 217
Mansfield Pass, Texas, 203, 204
Marco Island, Florida, 111
Matacumbe Key, Florida, 108
Matagorda Island, Texas, 29
Matagorda Peninsula, Texas, 179, 191
Mediterranean Sea, 9
Mexico Beach, Florida, 80
Miami Oolite, 105–107
microtidal 103
Midnight Pass, Florida, 44, 119
Miocene, 117, 118
Mississippi delta, 82, 94, 104, 166–168
Mississippi Sound, 163
mixed-energy inlet, 44
Mobile Bay, Alabama, 143, 146
morphodynamics, 21
mud, 47
Mustang Island, Texas, 188, 194
Mustang Island, Texas, 39, 188, 192, 194–197

Naples, Florida, 111
neap tide, 16
nearshore zone, 21, 22
New Pass, Florida, 44
New Zealand, 42
North Bunces Key, Florida, 123, 125
North Ca;tiva Island, Florida 113
nourishment, 85, 120–123, 126, 127, 158, 206

oblique bar, 23
Ocean Springs, Mississippi, 154
offshore breakwater, 175
ooids, 57, 58
Orange Beach, Alabama, 143, 145, 146
Oregon, 16, 18, 25

Pacific Ocean, 9
Packery Channel, Texas, 195, 198
Padre Island, 28, 37, 38, 179, 198-201, 205
Pascagoula. Mississippi, 154
Pass Christian, Mississippi, 155, 157
Pass-a-Grille, Florida, 123
peat, 128
pebble, 47
perched beach, 86, 87
Perdido Key, Alabama, 143, 145
Petit Bois Island, Mississippi, 157, 159
petroleum industry, 169
phi unit(Φ), 48
phosphate, 117, 118
    shark teeth, 118
Pleistocene, 221, 222, 223
plunging wave, 10, 11
Point of Rocks, Florida, 33, 119, 120
Port Aransas, Texas, 188, 189, 190, 195
Port Mansfield, Texas, 78, 199
Portuguese Man-of-war, 94, 95
post-storm beach, 24
progradation, 37

quartz, 53–58
Quintana Roo, Mexico, 216

railroad vine, 98
red tide, 100–102
reefs, 221, 223
rhythmic topography, 29, 30
ridge and runnel, 24–26
rip current, 14–16, 22
rip current channel, 23
ripples, 59-61
riprap, 70, 71
rock fragments, 54, 58
rocky coast, 34
roundness, 51, 53, 54

Sabine River, 180
saddle, 14, 22
salient, 72, 169
San José Island, Texas, 188, 192, 193
sand, 47, 128
    sand bodies, 83
sand dollar, 91, 92
Sand Key, Florida, 70, 73- 75, 86, 87, 127
sand shadows, 62
Sanibel Island, Florida, 112
Sargassum, 88, 101, 102, 108, 191, 198, 199, 203, 204, 205

Sargent Beach, Texas, 31, 33, 50, 67, 187, 189, 190
Scandanavia, 17
sea level, 17
sea oats, 101
sea turtle, 95- 98, 214
sea-level rise, 20, 167
sea-level change, 17, 19
seawall, 21, 69, 70, 72, 128, 185, 220
sediment budget, 14
sedimentary structures, 59
    antidunes, 60, 61
    burrow, 94, 97, 99
    current crescents, 62
    ripples, 59- 61
    sand shadows, 62
sediment-starved coast, 82
semi-diurnal tide, 16
shark teeth, 118
sheet piling, 75
shell debris, 82
shell gravel, 202
Ship Island, Mississippi, 153, 160, 162–164
shoreline change, 83
Siesta Key, Florida, 33, 34, 119, 120, 122
silt, 47
simulation model, 6
snail, 91
solar tide, 16
sorting, 48
South Padre Island, Texas, 88, 89, 104, 203, 206, 207, 208
sphericity, 51
spilling wave, 10,11
sting ray, 91, 92
storm drain, 157
storms, 31
    storm surge, 7

suction dredge, 82, 84
supratidal, 35
surf clam, 92, 93
surf fishermen, 198
surf zone, 12, 85, 91
Surfside, Texas, 188
Surfside, Texas, 89
surging wave, 10,11
swash marks, 61
swash zone, 26, 94, 203, 220, 221
swell waves, 12

Tabasco, Mexico, 212, 213, 214
Tahiti, 49
Tamaulipas, Mexico, 209, 210, 211
Tampa Bay, Florida 120, 124, 125
Ten-Thousand Islands, Florida, 105
terminal groin, 75, 77, 189
terminal lobe, 45
terrigenous, 54, 209, 213
Tertiary, 35
Texas coast, 4, 7
tidal inlet, 36, 37, 42, 43, 82, 199
    Big Sarasota Pass, Florida, 45, 120
    tide-dominated inlet, 42, 45
    Venice Inlet, 118
tide-dominated inlet, 42, 45
tides, 14
    tidal prism, 37, 40
    tidal range, 24, 41, 42
tie-downs, 72
trade winds, 3, 7
transfer station, 81
transgressive, 37
transverse bars, 24
Treasure Island, Florida, 77, 125, 126
tropical storm, 7
Tropical Storm Dolly 2008, 206

troughs, 21
Turtle Beach, Florida, 120
turtles, 214
    turtle eggs, 96, 97
    turtle nest, 126

Upham Beach, Florida, 124, 125

Varadero Beach, Cuba, 104, 220, 223
Venice Inlet, 118
Venice, Florida, 117, 118
Veracruz, Mexico, 35, 212, 213

Washington coast, 16
washover apron, 31, 32
washover channels, 186, 187
washover fan, 31, 39
waves, 9
    wave energy, 69, 183
    wave length, 9, 29
    wave period, 9
    wave refraction, 12, 13
    wave runup, 62
wave-dominated coast, 179
    wave-dominated barrier, 38, 119
    wave-dominated delta, 187
    wave-dominated inlet, 44
    wave-generated currents, 88
Waveland, Mississippi, 158

Yarborough Pass, Texas, 199, 200
Yucatan Peninsula, 35, 58, 209
Yucatan, Mexico, 216, 218

Other books in the Harte Research Institute for Gulf of Mexico Study Series:

Coral Reefs of the Southern Gulf of Mexico — Tunnell, Chávez, and Withers

Gulf of Mexico Origin, Waters, and Biota: Volume 1, Biodiversity — Felder and Camp

Gulf of Mexico Origin, Waters, and Biota: Volume 2, Ocean and Coastal Economy — James C. Cato

Encyclopedia of Texas Seashells — Tunnell, Andrews, Barrera, and Moretzsohn

Sea-Level Change in the Gulf of Mexico — Richard A. Davis Jr.

Gulf of Mexico Origin, Waters, and Biota: Volume 3, Geology — Buster and Holmes

Arrecifes Coralinos del sur del Golfo de México — Tunnell, Chávez, and Withers

Gulf of Mexico Origin, Waters, and Biota: Volume 4, Ecosystem-Based Management — Day and Yáñez-Arancibia